THE FUTURE OF THE BRAIN

THE
FUTURE
OF THE BRAIN

ESSAYS BY THE WORLD'S
LEADING NEUROSCIENTISTS

EDITED BY
GARY MARCUS AND JEREMY FREEMAN

PRINCETON UNIVERSITY PRESS
PRINCETON AND OXFORD

Requests for permission to reproduce material from this work should be sent to Permissions,
Princeton University Press
Published by Princeton University Press, 41 William Street, Princeton, New Jersey 08540
In the United Kingdom: Princeton University Press, 6 Oxford Street, Woodstock, Oxfordshire
OX20 1TW

press.princeton.edu

Jacket design by Karl Spurzem

ISBN 978-0-691-16276-8

Library of Congress Control Number: 2014938489

British Library Cataloging-in-Publication Data is available

This book has been composed in Minion Pro

Printed on acid-free paper. ∞

Printed in the United States of America

10 9 8 7 6 5 4 3 2 1

CONTENTS

IMPLICATIONS

AFTERWORD

CONTRIBUTORS

Misha B. Ahrens
HHMI, Janelia Farm Research Campus, Ashburn, Virginia

Ned Block
New York University

Arthur Caplan
New York University, Langone Medical Center

Matteo Carandini
University College London

George Church
Harvard Medical School

John Donoghue
Brown Institute for Brain Science, Brown University

Chris Eliasmith
University of Waterloo, Ontario, Canada

Simon E. Fisher
*Max Planck Institute for Psycholinguistics and Donders Institute for Brain,
Cognition & Behaviour, Radboud University, Nijmegen, the Netherlands*

Jeremy Freeman
HHMI, Janelia Farm Research Campus, Ashburn, Virginia

Mike Hawrylycz
Allen Institute for Brain Science, Seattle, Washington

Sean Hill
Brain Mind Institute, École Polytechnique Fédérale de Lausanne, Switzerland

Christof Koch
Allen Institute for Brain Science, Seattle, Washington

Leah Krubitzer
University of California, Davis

Michel M. Maharbiz
University of California, Berkeley

Gary Marcus
New York University and Allen Institute for Brain Science, Seattle, Washington

Kevin J. Mitchell
Smurfit Institute of Genetics and Institute of Neuroscience, Trinity College, Dublin

May-Britt Moser and Edvard I. Moser
Centre for Neural Computation, Kavli Institute for Systems Neuroscience, Norwegian University of Science and Technology, Trondheim, Norway

David Poeppel
New York University

Krishna V. Shenoy
Departments of Electrical Engineering, Bioengineering, and Neurobiology, Stanford University

Olaf Sporns
Indiana University, Bloomington

Anthony Zador
Cold Spring Harbor Laboratory, New York

PREFACE

There's never been a more exciting moment in neuroscience than now. Although the field has existed for two centuries, going back to the days of Phineas Gage and the tamping iron that exploded through his left frontal lobe, progress has in many ways been slow. At present, neuroscience is a collection of facts, still awaiting an overarching theory; if there has been plenty of progress, there is even more that we don't know. But a confluence of new technologies, many described in this book, may soon change that.

To be sure, there is long history of advances, even from the earliest days, often leveraging remarkably crude tools to great effect. In the mid-1800s Paul Broca got the first glimpse into the underpinnings of language by doing autopsies on people who had lost linguistic function because of brain damage to specific cortical areas. Near the end of the nineteenth century, Camillo Golgi discovered that he could visualize neurons under a microscope by staining them with silver nitrate, and Santiago Ramón y Cajal used the technique to develop remarkably prescient characterizations of neuronal structure and function. In 1909 a brilliant ophthalmologist named Tatsuji Inouye launched functional brain mapping, by methodically studying victims of gunshot wounds during the Russo-Japanese war, noting that wounds to the visual cortex impaired his patients' vision, and wounds to particular locations affected vision in *particular* regions of the visual field.

In the latter part of the twentieth century, noninvasive forms of brain imaging, like functional magnetic resonance imaging (fMRI), came on the scene. But as useful as such tools are, current noninvasive techniques are like fuzzy microscopes; they blur the fine detail of neural activity in both space and time. Ultimately, looking at an fMRI scan is like looking at a tiny pixelated version of a detailed, high-resolution photograph.

In nonhuman animals, which can be studied with more invasive techniques, the gold standard until recently was the "single neuron recording," which uses thin electrodes to monitor the electrical activity associated with neural firing. Action potentials are the currency of the brain, and directly measuring them has led to many fundamental insights, such as Hubel and Weisel's discovery that neurons in the visual cortex are "tuned" or selective for particular visual features. But looking at one neuron at a time tells an incomplete story at best; the neuroscientist Rafael Yuste has likened it to "understanding a television program by looking at a single pixel."

As we write this, it is clear that neuroscience is undergoing a revolution. Optogenetics, introduced in 2005, makes it possible to engineer neurons that literally light up when active, switching them on and off with a laser; multielectrode recordings, which allow recordings from hundreds or even thousands of neurons are finally becoming practical, and new forms of microscopy can record the activity of nearly every neuron in a living, transparent fish. For the first time, it is realistic to think that we might observe the brain at the level of its elementary parts.

● ● ● ● ● ●

Still, three fundamental truths make the brain more challenging to understand than any other biological system.

First is sheer numbers. Even in the fly or the larval zebrafish brain there are one hundred thousand neurons. In the human brain there are over 85 billion. On top of that, the word *neuron* makes it sound like there is only one kind, whereas in fact there are several hundred kinds, possibly more, each with distinctive physical characteristics, electrical characteristics, and, likely, computational functions. Second, we have yet to discover many of the organizing principles that govern all that complexity. We don't know, for example, if the brain uses anything as systematic as, say, the widespread ASCII encoding scheme that computers use for encoding words. And we are still shaky on fundamentals like how the brain stores memories and sequences events over time. Third, many of the behaviors that seem characteristically human—like language, reasoning, and the acquisition of complex culture—don't have straightforward animal models.

The Obama BRAIN Initiative, the European Human Brain Project, and other large-scale programs that may begin in Asia aim to address some of the challenges in understanding the brain. It seems reasonable that we can expect, over the next decade, an enormous amount of new data at an unprecedented level of detail, certainly in animals, and perhaps in humans as well. But these new data will raise new questions of their own. How can researchers possibly make sense of the expected onslaught of data? How will we be able to derive general principles?

And for that matter, will collecting all these data be enough? How can we scale up data analysis to the terabytes to come, and how can we build a bridge from data to genuine insight? We suggest that one key focus must be on computation. The brain is not a laptop, but presumably it is an information processor of some kind, taking in inputs from the world and transforming them into models of the world and instructions to the motor systems that control our bodies and our voices. Although many neuroscientists might take for granted that the principal process by which the brain does its work is some form of computation, almost all agree that the most foundational properties of neural computation have yet to be discovered. Our hope is that computation can provide a universal language for describing the action of the brain, especially as theorists and experimentalists come closer together in their quest.

Given the complexity of the brain, there is no certainty we will come to fully or even largely understand the brain's dynamics anytime soon; in truth, there is reason for hope, but no guarantees. This book, with chapters by pioneers like Christof Koch and George Church, represents our best guesses—and our esteemed contributors' best guesses—about where we are going, what we are likely to find out, and how we might get there.

But it also admits where we might stumble along the way. If this book is a reader's guide to the future, it's not a foolproof crystal ball; if anything, it's more like a time capsule. Part of the fun will be for scientists, policy makers, and the public to come back to these essays a decade hence, to, as one colleague put it, "reassess its scientific claims, aspirations, and methodological promises, and adjust the aspirations of the next generation of neuroscientific endeavors accordingly." We couldn't agree more.

Gary Marcus and Jeremy Freeman
March 2014

MAPPING THE BRAIN

In the jargon of neuroscience, to map the brain is to understand two things: all of the brain's myriad connections (equivalent to drawing a map of all the roads and buildings in the United States) and all of the "traffic" (neural activity that occurs on those roads). "Connectomes" are like highway maps, "activity maps" record the traffic as the brain is engaged in behavior. Like Google Maps, we ultimately need many "layers" of information, telling us about landmarks (like the folds of the cortex), annotations about particular types of neurons (the brain likely has close to a thousand), and ultimately about the pathways of neurons that are involved in particular kinds of behaviors.

The essays in this part tell a story—from the current, cutting edge to the future—about technological advances that will allow us to map out as much of that territory as possible. Most complex organisms have hundreds of thousands, if not millions or billions, of neurons. For decades, neuroscientists have recorded from just a few at a time, inferring something about a complex system based on incomplete measurements. **Mike Hawrylycz** narrates the history of brain anatomy, from the earliest drawings of neural circuits by Ramón y Cajal to ongoing, cutting-edge efforts to obtain and annotate high-resolution anatomical maps of the entire human brain at cellular resolution. **Misha Ahrens** describes an approach called light-sheet microscopy for monitoring neural activity from the *entire* brain of a transparent organism, the zebrafish, and to do so during behavior in intact animals. **Christof Koch** describes a confluence of emerging methods—anatomical, physiological, and optical—that are making it possible to characterize neural activity across large swaths of the visual cortex of the mouse. Looking further into the future, **Anthony Zador** and **George Church** describe novel approaches to characterizing neural anatomy, specifically neural connectivity, that use genetic techniques to indirectly encode information about connectivity in sequences of DNA. **Church** discusses how these approaches might even be extended to record the firing of neurons over scales much larger than optical or electrophysiological methods currently allow.

BUILDING ATLASES OF THE BRAIN

Mike Hawrylycz

With Chinh Dang, Christof Koch, and Hongkui Zeng

A Very Brief History of Brain Atlases

The earliest known significant works on human anatomy were collected by the Greek physician Claudius Galen around 200 BCE. This ancient corpus remained the dominant viewpoint through the Middle Ages until the classic work *De humani corporis fabrica* (*On the Fabric of the Human Body*) by Andreas Vesalius of Padua (1514–1564), the first modern anatomist. Even today many of Vesalius's drawings are astonishing to study and are largely accurate. For nearly two centuries scholars have recognized that the brain is compartmentalized into distinct regions, and this organization is preserved throughout mammals in general. However, comprehending the structural organization and function of the nervous system remains one of the primary challenges in neuroscience. To analyze and record their findings neuroanatomists develop atlases or maps of the brain similar to those cartographers produce.

The state of our understanding today of an integrated plan of brain function remains incomplete. Rather than indicating a lack of effort, this observation highlights the profound complexity and interconnectivity of all but the simplest neural structures. Laying the foundation of cellular neuroscience, Santiago Ramón y Cajal (1852–1934) drew and classified many types of neurons and speculated that the brain consists of an interconnected network of distinct neurons, as opposed to a more continuous web. While brain tissue is only semitranslucent, obscuring neuronal level resolution, a certain histological stain Franz Nissl (1860–1919) discovered, and known as the *Nissl* stain, can be used to stain negatively charged RNA in the cell nucleus in blue or other visible colors. The development of this stain allowed the German neuroanatomist

a

ANDREAE VESALII
BRVXELLENSIS, DE HVMANI CORPO-
RIS FABRICA LIBER SEPTIMVS, CEREBRO ANL
malis facultatis fedi & fenfuum organis dedicatus, & mox in initio omnes
propemodum ipfius figuras, uti & duo proximè præceden
tes libri, commonftrans.

PRIMA SEPTIMI LIBRI FIGVRA·

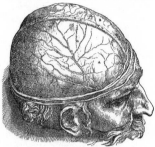

b

QVINTA SEPTIMI LIBRI FIGVRA·

PRÆSENS figura quòd ad reliftam in caluaria cerebri portionê attinet, nullà ex parte uariat:atq; id folù habet proprium, quod callofum corpus hic anteriori fua fede à cerebro primùm liberauimus, ac dein eleuatum in pofteriori refleximus, feptum dextri ac finiftri uentriculorum dì uellentes, & corporis inftar teftudinis extructi fuperiorem fuperficiem ob oculos ponètes. Ab A A, A, A itaq; & B, B, ad Q B,ac dein D,D, D, & E & F, & G & H eadem hic indicant,quæ in quartafigura.Sic quoque & L,L,& M, M,& O & P & Q eadem infinuant.

R,R, R Notatur inferior callofi corporis fuperficies.eft enim id à fua fede motum,atque in pofteriora reflexum.

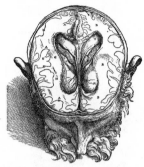

S,T,V Supe-

Figure 1. a. Cover of the work *De humani Corporis fabrica libri septem*, published by Andreas Vesalius in 1543. The work was the first major advance in human anatomy since the Greek physician Galen. b. A page from the fifth chapter of the book showing the cortex and ventricles of the brain.

Korbinian Brodmann (1868–1918) to identify forty-three distinct regions of the human cerebral cortex based on cytoarchitectural organization using this Nissl stain. These pioneering works of Brodmann, Constantin von Economo, Marthe Vogt, and others mapped cyto- and myeloarchitectural landscape of the human cortex based on painstaking visual inspection and characterization of a few observable cellular properties such as cell shape, density, packing, and such.

Since Vesalius, most atlases of the brain have been drawn on paper, with the most recent versions in vivid color delineating hundreds of structures. Such atlases have been drawn for most of the important model organisms studied in the laboratory and provide key bench-side experimental references. As with most aspects of modern biology, however, technology has been a driving factor in improved understanding of brain organization. Neuroimaging techniques evolved over the last twenty years have now allowed neuroscientists to revisit the subject of brain mapping, with the modern brain atlas more akin to a digital database that can capture the spatiotemporal distribution of a multitude of physiological and anatomical data. Modern techniques such as magnetic resonance imaging (MRI), functional magnetic resonance imaging (fMRI), diffusion MRI, magnetoencephalography (MEG), electroencephalography (EEG), and positron emission tomography (PET) have provided dramatic improvements in brain imaging for research, clinical diagnosis, and surgery. Digital atlases based on these techniques are advantageous since they can be *warped*, mathematically or *in silico*, to fit each individual brain's unique anatomy.

The origin of modern brain mapping for clinical use lies with the seminal work of Jean Talairach, who in 1967 developed a 3D coordinate space to assist deep brain surgical methods. This atlas was generated from two series of sections from a single sixty-year-old female brain, and was later updated by Talairach and P. Tournoux in a printed atlas design for guiding surgery. Today biomedical imaging forms a crucial part of diagnosis and presurgical planning, and much time and resources are invested in the search of imaging biomarkers for diseases. Atlases have been used in image-guided neurosurgery to help plan "stereotaxic," that is, coordinate referenced, neurosurgical procedures. Using this data, surgeons are able to interpret patient-specific image volumes for anatomical, functional, and vascular relevance as well as their relationships.

The field of digital atlasing is extensive and includes high-quality brain atlases of the mouse, rat, rhesus macaque, human, and other model organisms. In addition to atlases based on histology, magnetic resonance imaging, and positron emission tomography, modern digital atlases use gene expression, connectivity, and probabilistic and multimodal techniques, as well as sophisticated visualization software. More recently, with the work of Alan Evans at the Montreal Neurological Institute and colleagues, averaged standards were created such as the Colin27, a multiple scan of a single young man, as well as the highly accessed MNI152 standard. While inherently preserving the 3D geometry of the brain, imaging modalities such as MRI, CT, and PET do not usually allow for detailed analysis of certain structures in the brain because of limitations in spatial resolution. For this reason it is common to use very high-resolution 2D imaging of *in vitro* tissue sections and employ mathematically sophisticated reconstruction algorithms to place these sections back into the 3D context of the brain.

Today digital brain atlases are used in neuroscience to characterize the spatial organization of neuronal structures, for planning and guidance during neurosurgery, and as a reference for interpreting other data

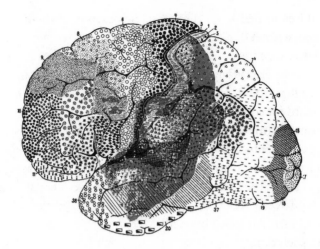

Figure 2. Regions of the human cerebral cortex delineated by Korbinian Brodmann using Nissl stain histology. Brodmann identified forty-three distinct regions that today still serve as a guide for studying distinct functional areas in the human cortex.

modalities such as gene expression or proteomic data. One ultimate aim of neuroscientific inquiry is to gain an understanding of the brain and how its workings relate to activities from behavior to consciousness. Toward this end, brain atlases form a common coordinate framework for summarizing, accessing, and organizing this knowledge and will undoubtedly remain a critical-path technology in the future.

The Genetic Brain

The development of the techniques of modern molecular biology and eventually whole genome sequencing opened the door for understanding the genetics of the brain, and new perspectives on the study of brain anatomy are emerging with the availability of large-scale spatial gene expression data. The brain consists of at least several hundred distinct cell types whose complete classification is still at present elusive. Each cell type is related to its function with its gene expression pattern, for example, on/off, high/low, as a key determinant. Gene expression data can be collected through a variety of techniques, and exploration of these data promises to deliver new insights into the understanding of relations between genes and brain structure.

Early gene expression studies used methods such as northern blots, which combine electrophoresis separation of RNA molecules followed by hybridizing probes for detection. At one time this method was the gold standard for confirming gene expression, but it ultimately gave way to more quantitative methods. The microarray revolution dramatically increased our ability to profile genes by hybridizing many gene probes on a single gene chip. Today rapid digital sequencing technology can count individual RNA fragments that can subsequently be mapped back to the genome once it is known for an organism.

In 2001, Paul Allen, cofounder of Microsoft, assembled a group of scientists, including James Watson of Cold Spring Harbor Laboratory and Steven Pinker, then at MIT, to discuss the future of neuroscience and what could be done to accelerate neuroscience research. During these meetings the idea emerged that a complete 3D atlas of gene expression in the mouse brain would be of great use to the neuroscience community. The mouse was chosen due to the wealth of existing genetic studies and for practical reasons. Of the potential possible techniques, the project chose a technique for mapping gene expression called *in situ* hybridization (ISH) (automated by Gregor Eichele of the Max Planck Institute and colleagues), which uses probes that bind to mRNA within sectioned but intact brain tissue and thereby preserves spatial context (see color plate 1).

In 2006, an interdisciplinary scientific team at the Allen Institute for Brain Science, funded by Paul Allen and led by Allan Jones, delivered the first atlas of gene expression in a complete mammalian brain, publically available online at www.brain-map.org. Since then, the Allen Institute has expanded its projects to provide online public resources that integrate extensive gene expression, connectivity data, and neuroanatomical information with powerful search and viewing tools for the adult and developing brain in mouse, human, and nonhuman primate (see figure 3 for an example). In addition to the data there are colorimetric and fluorescent ISH image viewers, graphical displays of ISH, microarray and RNA sequencing data, and an interactive reference atlas viewer ("Brain Explorer") that enables 3D navigation of anatomy and

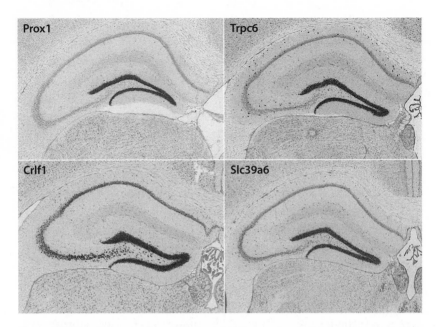

Figure 3. Genes whose expression pattern is highly correlated with *Prox1* (upper left) in dentate gyrus of the hippocampus. These genes were found by starting with the image for gene *Prox1* and searching for patterns whose spatial pattern of gene expression strongly resembled *Prox1*. Combinations of expression patterns such as these may help to refine our present understanding of the function of the hippocampus.

gene expression across these datasets. (Approximately fifty thousand users worldwide access the Allen Brain Atlas resources each month.) Scientists have mined the atlases to search for marker genes in various brain regions associated with diseases, to identify different cell type markers, to delineate brain regions, and to compare gene expression data across species.

Extending this work to humans, the Allen Human Brain Atlas was made public in May 2010 and is the first anatomically comprehensive and genome-wide, three-dimensional map of the human brain. This transcriptional atlas of six adult human brains contains extensive histological analysis and comprehensive microarray profiling of several hundred precise brain subdivisions and has revealed that gene expression varies enormously by anatomical location, with different regions and their constituent cell types displaying robust molecular signatures that are highly conserved between individuals.

In particular, these data show that 84 percent of all genes are expressed somewhere in the human brain and in patterns that while complex are substantially similar from one brain to the next. The analysis of differential gene expression and gene coexpression relationships demonstrates that brain-wide variation strongly reflects the distributions of the major cell types such as neurons, oligodendrocytes, astrocytes, and microglia, all of which are essential to brain function. Interestingly, the neocortex displays a relatively homogeneous transcriptional pattern but with distinct features associated selectively with primary sensorimotor cortices and with enriched frontal lobe expression. Interestingly, the spatial topography of the neocortex is strongly reflected in its molecular topography, that is, the closer two cortical regions are, the more similar their gene expression patterns remain.

Several other significant efforts toward understanding the genetic basis of brain organization are underway, including the Edinburgh Mouse Atlas Project (EMAP) (www.emouseatlas.org), which contains substantial spatial and temporal data for mouse embryonic development, and the Rockefeller University–based GENSAT project of Nathanial Heintz and colleagues that seeks to characterize gene expression patterns using Bacterial Artificial Chromosomes (BAC) in genetically modified mice (www.gensat.org), as well as BGEM (www.stjudebgem .org), GenePaint (www.genepaint.org), EurExpress (www.eurexpress .org), and MGI (http://www.informatics.jax.org), all generally user friendly with useful tutorials.

A Standard Brain?

Does a standard or normal brain exist? This is less likely for humans than genetically bred mice, but mapping neuroscientific and clinical data onto a common frame of reference allows scientists and physicians to compare results between individuals. One main reason for standardization is that multiple and diverse brains can be transformed into a standard framework that maximizes our ability to understand their similar features. Another is that it allows us to identify how unique or unusual features in a particular brain may differ from an average population. With modern advanced image processing capabilities, digital atlases can serve as the framework for building standard atlases and for surveying the information linked to it. In contrast to basic data repositories, which allow for simple access to data through a single interface, sophisticated digital atlases backed by appropriate technology can act as hubs facilitating access to multiple databases, information sources, and related documents and annotations. These may act as a scaffold from which to share, visualize, analyze, and mine data of multiple modalities, scales, and dimensions.

Many of these ideas of standardization grew out of a major initiative of the National Institutes of Health in the 1990s called the "Decade of the Brain," where a number of digital and electronic resources were created to enable the unification and integration of the various subfields of neuroscience. One outcome of this work is the field of *neuroinformatics*, or the application of computer- and mathematical-based technologies to organize and understand brain data. The ultimate goal of neuroinformatics is to bring together brain architecture, gene expression, and 2D and 3D imaging information into common frames of reference. Major organizations have evolved around mapping brain data, such as the International Consortium for Human Brain Mapping (www.loni.ucla .edu/ICBM/About) and the International Neuroinformatics Coordinating Facility (INCF, www.incf.org). These efforts have led to atlases such as the standard Talairach Atlas and the Montreal Neurological Institute (MNI) standard that have been extensively used in neuroscience.

One consideration in standardizing brain atlases is the type of coordinate system used. As Alan Evans of the Montreal Neurological Institute remarks, "The core concept within the field of brain mapping is the use of a standardized or 'stereotaxic' 3D coordinate framework for data analysis and reporting of findings from neuroimaging experiments. This simple construct allows brain researchers to combine data from many subjects such that group-averaged signals, be they structural or functional, can be detected above the background noise." The concept of a coordinate system is fundamental to digital atlases and requires two basic components: the specification of an origin in the stereotaxic space and a mapping function that transforms each 3D brain from its native coordinates to that of the atlas. A major step in addressing these issues, and a standard tool set that allows different types of neuroscience data to be combined and compared, is now in development for one of the most important subjects in experimental neuroscience: the mouse, *Mus musculus*. This project is an international collaboration in digital atlasing and is sponsored in part by the International Neuroinformatics Coordinating Facility (INCF).

The Connected Brain

Much recent evidence indicates that the vast interconnected network of the human brain is responsible for our advanced cognitive capabilities, rather than a simple expansion of specialized regions of the brain such as the prefrontal cortex. This may apply in particular to diseases associated with potentially aberrant wiring such as schizophrenia, autism, and dyslexia. The importance of *circuit* considerations for differentially characterizing disorders such as major depression, anxiety, and obsessive-compulsive disorder, and substance abuses including nicotine addiction, is now being widely recognized.

It is now understood that neuropsychiatric disorders likely result from pathologies at the system level, with both complex genetic and environmental factors impacting neural circuitry. As Jason Bohland and colleagues point out in a recent 2009 proposal for a "mesoscale," that is, medium scale, connectome in *PLOS Computational Biology*: "For those [diseases] with heritable susceptibility effects, genetic polymorphism and cellular processes play a greater role, but anatomical circuits remain critical to understanding symptoms and developing therapies." In Parkinson's disease, for example, drug- and stimulation-based therapeutic interventions do not occur at a particular cellular lesion site, but rather are contingent on understanding interactions within the extrapyramidal motor system of neurons.

The first unified approach to defining connectional atlases of the brain was proposed by Olaf Sporns (Indiana University) and Patrick Hagmann (Lausanne). In 2005 they independently suggested the term "connectome" to refer to a complete map of the neural connections within the brain. This term was directly inspired by the concurrently ongoing effort to sequence the human genetic code, and since then the field of *connectomics* (see chapters by Sporns and Zador, this volume) has been concerned with assembling and analyzing connectome datasets. (The term *connectome* was most recently popularized in Sebastian Seung's book *Connectome*, which discusses the high-level goals of mapping the human connectome, as well as ongoing efforts to build a 3D neural map of brain tissue at the microscale.)

The first complete neural circuit in any organism was found in the common worm *Caenorhabditis elegans*, and research into its molecular and developmental biology was begun in 1974 by Nobel laureate Sydney Brenner. *C. Elegans* has since been used extensively as a model organism in biology. Using high-resolution electron microscopy and manual annotation of hundreds of images, the circuit-mapping project was a major tour de force of neuroanatomy, resulting in a 341-page publication by the Royal Society in 1986 by John White and Brenner titled "The Structure of the Nervous System of the Nematode *Caenorhabditis elegans*." Other landmark studies include a study of the areas and connections of the visual cortex of the macaque published by Daniel Felleman and David Van Essen in 1991 and of the thalamo-cortical system in the feline brain by J. W. Scannel and colleagues in 1999. Since then several neuroinformatics databases of connectivity have emerged, such as the online macaque cortex connectivity tool CoCoMac (www.cocomac.org) and the Brain Architecture Management System (BAMS, http://brancusi .usc.edu).

Several years ago, supported both by public and private funding, a series of independent projects were launched to map the connectome of the laboratory mouse at the mesoscale. Among these projects the Allen Institute embarked on a large-scale effort to develop a regional and cell type specific three-dimensional connectivity map. This Allen Mouse Brain Connectivity Atlas uses a combination of normal and genetically modified mice together with genetic tracing approaches and a high-throughput serial 2-photon tomography system to image the labeled axons throughout the entire brain. High-resolution coronal images are sampled every 100 μm (0.1 mm), resulting in a large 750-GB dataset per brain. At the end of 2013, approximately 1,500 terabytes of data (or 1.5 petabytes) will have been generated, all mapped onto a common 3D reference space of high spatial fidelity that allows for identification of the neural circuitry that governs behavior and brain function.

Mapping the connectome of the human brain is one of the great scientific challenges of the twenty-first century. The Human Connectome

Project (HCP, http://www.humanconnectome.org) is tackling a key aspect of this challenge by elucidating some of the main neural pathways that underlie brain function and behavior. Due to the immense complexity and comparatively large size of the human brain, the HCP (see chapter by Sporns, this volume) is taking a more macro approach to mapping large-scale circuitry, comprehensively mapping human brain circuitry in a target number of 1,200 healthy adults using a combination of noninvasive neuroimaging techniques such as MRI, EEG, and fMRI.

Accurate *parcellation* of fMRI imaging activity into component areas of the brain is an important consideration in deciphering its connectivity, and it takes us back to our original discussion of anatomy. Modern imaging techniques have enabled parcellation of localized areas of cortex and have been accomplished by using diffusion tractography and functional imaging to measure connectivity patterns and define cortical areas based on these different connectivity patterns. Such analyses may best be done on a whole brain scale and by integrating types of noninvasive imaging. It is hoped that more accurate whole brain parcellation may lead to more accurate macroscale connectomes for the normal brain, which can then be compared to disease states. The HCP images and their parcellations are being made available to the public through a public interface called the ConnectomeDB at http://www.humanconnectome.org, mentioned above.

The current noninvasive imaging techniques cannot capture the brain's activity at a neuronal level, and mapping the connectome at a cellular level in vertebrates currently requires postmortem microscopic analysis of limited portions of brain tissue. The challenge of doing this on a grand scale is quite major, as the number of neurons comprising the brain easily ranges into the billions in more highly evolved organisms, with the human cerebral cortex alone contains at least 10^{10} neurons and linked by 10^{14} synaptic connections. A few of the main challenges of building a microscale mammalian connectome today include: the data collection would take years given current technology; annotation tools are insufficient to fully delineate and extract information at a neuronal scale; and, not least, the algorithms necessary to map relevant connections and build the connectivity graphs are not yet fully developed. To address these machine-vision and image-processing issues, the Open

Connectome Project (openconnectome.org) is a crowd-sourcing initiative to meet this challenge. Finally, statistical graph theory is an emerging discipline that is developing sophisticated pattern recognition and inference tools to parse these brain graphs.

The Future Brain

Development of large-scale brain atlases is now a major undertaking in neuroscience. While it may not be possible to systematically map each of the one hundred billion neurons any time soon in any given individual brain, modern mapping techniques are providing atlases of remarkable resolution and functionality.

Several recent advances in neuroimaging support the possibility of deep and large-scale mapping, and this goal may be less audacious than seems at first. For example, using a combinatorial color labeling method, Brainbow, which is based on the random expression of several types of fluorescent proteins, Josh Sanes and Jeff Lichtman at Harvard are able to mark individual neurons with one of over one hundred distinct colors. The labeling of individual neurons with a distinguishable hue then allows the tracing and reconstruction of their cellular structure, including long processes within a block of tissue. Labeling techniques such as these allow for classification and visualization of microscopic neurons. Another approach aimed at classifying diversity in the synaptic code, called array tomography, has been developed by Stephen Smith at Stanford, and can also achieve combinatorial labeling of synaptic connections using electron microscopy.

Recently, in a processing tour de force, nearly 7,500 sections of an individual human brain were sliced and scanned and mapped onto a 3D reconstructed brain at 20-micron isotropic resolution, that is, in all three spatial dimensions. This project is the culmination of years of work from the Katrin Amunts and Karl Zilles laboratory at Jülich, Germany, with semiautomated informatics reconstruction by Evans in Montreal. The atlas called BigBrain is a thin-sliced histology project that offers nearly cellular resolution, that is, detail close to the level of the cell. Because of the nearly continuously collected sections and the 3D image reconstruction, BigBrain is a dataset 125,000 times bigger than a typical MRI! Atlases based on MRIs do not allow for the visualization of information at the level of cortical cells and layers, although this atlas will allow that. However, to make BigBrain into a full-fledged atlas it will need to be annotated, that is, it will need to provide the anatomic structural delineations that outline the fine structure of the brain.

The BigBrain effort indicates that high-resolution 3D microscopy is still not at a level of resolution to map the finest structures in the brain. However, advances are being made in 3D imaging as well. In 2013, in a highly publicized article in *Nature*, a method was developed to subject the brain to a three-dimensional network of hydrophilic polymers and then to remove the lipids from the brain by electrophoresis. The brain remains fully intact but optically transparent and macromolecule permeable. Using mouse brains, the authors show intact-tissue imaging of long-range projections, local circuit wiring, cellular relationships, and subcellular structures. This method, called "CLARITY," uses intact-tissue in situ hybridization and immunohistochemistry with multiple rounds of staining and de-staining in nonsectioned tissue to visualize gene expression or protein binding. The method is still being refined but may be useful for human postmortem imaging as well.

Alternative computational processing techniques will also be necessary to deal with the massive data these new atlases generate. In 2012, a Citizen Science project called EyeWire, launched by Sebastian Seung of MIT, began attempting to crowd source the mapping of the connectome through an interactive game in which contributors try to map the retinal connectome (Zador's chapter herein outlines another possible approach to this problem).

Large-scale atlases of the brain are providing content to the neuroscience community through molecular, cellular, functional, and connectomic data. The transition from print to digital atlases has been revolutionary, as it has allowed navigation, 3D reconstruction, and visualization from the smallest nuclei to macroscale regions. Digital atlases have also transformed clinical neuroscience, and all stages from pre- to postoperation of surgery in some way use digital atlases. It is likely that in the near future we will have annotated 3D microscale atlases of the structure of the human brain. In several years it should be possible to achieve near complete axon and synaptic connectivity in a substantial segment of the human cortex, thereby elucidating the detailed complexity of its cortical circuitry. Atlases will continue to be more integrated into scientific and clinical workflows, thus aiding in discovery science and providing novel ways of diagnosing, monitoring, and treating disease.

WHOLE BRAIN NEUROIMAGING AND VIRTUAL REALITY

Misha B. Ahrens

Historically, the brain has been studied in small chunks, such as by recording from individual or small groups of cells, making it challenging to relate discoveries at the small network level to function that relies on whole-brain mechanisms. As this book goes to press, there are a number of efforts underway aimed at recording activity from ever-more neurons simultaneously, presumably an important step toward a fuller understanding of how the brain works and how the computations the many thousands or billions of its constituents perform make up the function of the whole (see chapter by Shenoy, this volume). In this essay, we lay out a strategy for imaging from virtually *all* the neurons in the brain of a vertebrate animal, the larval zebrafish.

Before explaining how it might be done, it is important to understand at the outset why even that—imaging from all neurons—is not enough for fully understanding brain function. Looking at brain activity in isolation ignores the physical *context* of this organ: the brain resides inside a body with its own dynamics, and the body, in turn, operates in a physical environment that follows the laws of nature. The neurons inside the brain are massively interconnected, but, critically, the output of the brain directly feeds back into it through the environment, because "decisions" the brain makes lead to actions that change the configuration of the body and the external environment and that elicit new sensory input, which in turn is processed by the brain, forming what is called the sensorimotor loop. This loop is important—for example, visual input resulting from a decision to start walking is processed differently from the same visual input in the absence of that decision. In this sense, understanding the brain might therefore best be seen as a holistic problem of understanding the entire dynamical system as a whole: brain, body, and environment.

In certain cases, we can develop a reasonable understanding of neural function without a holistic view, working strictly from the bottom up. For example, much progress has been made in understanding the retina, olfactory bulb, primary visual cortex, and the peripheral auditory system, to the extent that it is now possible, for example, to create cochlear implants that allow some deaf people to hear (see chapter by Donoghue herein). Other questions, which involve, for example, interactions between sensory input, memory and action, may best be addressed by studying all parts of the whole—all sensory input, activity of all neurons, all motor output, and the pathways from motor output back to sensory input.

In collaboration with Florian Engert, Philipp Keller, and others, building on work in insects and mammals and on developments in microscopy, we developed two experimental systems that are starting to allow for the holistic study of the dynamical system described above in a vertebrate species, the zebrafish. The first piece of technology creates an artificial context for the nervous system by means of virtual reality, facilitating neural recordings because the animal's head stays in place; the second gives us the ability to record from almost every neuron in the larval zebrafish brain—about 80,000 out of the total of 100,000 neurons. Taken together, the hope is that exhaustive measurement of context as well as of neural activity will allow insights into neural function that were hitherto unachievable.

Closing the Sensorimotor Loop

Many real-world behaviors, such as walking or flying, depend on real-time feedback and can be altered based on that feedback. In walking, for example, a stumbling organism may correct gait based on feedback to the vestibular and visual systems. To study the behavior holistically, this other part of the complete dynamical system—the relation between brain and the environment—has to somehow be incorporated into the experimental setup.

Since the 1960s, researchers have been incorporating such real-time feedback for the study of behavior in insects. Scientists like Karl Götz and Martin Heisenberg created such experimental systems for flies, which they glued to a thin wire and allowed to flap their wings. A signal from a sensitive torque meter that measures miniature rotational forces as the flies tried to steer left or right was used to rotate a drum so that a visual pattern moved in the opposite direction. In this way a stationary fly was given the realistic visual feedback it would have received were it freely flying. Essentially, this formed a simple virtual-reality ride for the fly. The goal of this research isn't just to fool animals into thinking they are freely behaving when they're not; it is to understand the details of their behavior and, ultimately, how the brain controls the animal. Therefore, more recently, researchers have taken this approach one step further and have started recording from brains of animals embedded in such closed-loop virtual-reality systems. With the heads of the animals stationary, the brain is relatively easily accessible to neural recordings, via microscopes and electrodes. In recent work by David Tank and others, mice run around in three-dimensional virtual-reality environments, while the activity of neurons representing their spatial location or decision processes is monitored.

In studying larval zebrafish, we took a slightly different approach, using paralyzed animals. In this preparation, the brains are entirely stationary, hence easy to record from or manipulate. How can one create a virtual-reality environment for a paralyzed animal, when the goal is to study how the body, brain, and the environment interact dynamically? The particular paralysis method we used affects only the connection between neurons and the muscles, but the central nervous system and the spinal cord still function: the neural commands to the tail muscles are still intact. We can record from these neurons and interpret their activity as reflecting the animal's intent, akin to how the characters in the movie *The Matrix* moved through their virtual world via recordings from their brains.

When passed through an audio amplifier, the neural recordings from the fish's tail have a characteristic crackling sound in the rhythm of the tail undulations of freely swimming fish. These signals could be converted into virtual swim bouts that propelled the animals through a virtual reality environment displayed underneath them using a video projector. As such, we were now in a position to record from the brains of these animals during behavior. The next question is what brain area to record from, since the generation of behavior relies on sensory

systems, motor systems, and everything in between. In the larval zebrafish, our approach was to try and record from almost all neurons at
the same time.

Whole-Brain Neuron Imaging

The larval zebrafish has a number of experimental advantages that
neuroscientists have recently started to exploit and that eventually allowed us to build a microscope to record from almost all of its neurons.
At a young larval stage, about a week old, these animals actively swim
around, performing behaviors such as exploration, food seeking, and
simple forms of learning. Their brains are small, consisting of about
100,000 neurons, which presumably makes it an easier system to understand than large mammalian brains. Moreover, they share much of their
brain architecture with that of humans. Perhaps most importantly, they
are transparent, and certain mutants lack pigment in the skin, which allows light microscopes to penetrate down to the deepest layers of their
brain, so that essentially all neurons are accessible for imaging. Furthermore, zebrafish are a genetic model organism and can be made to express genetically engineered proteins in all or subsets of neurons. In the
last two decades, scientists have created proteins that report neural activity through calcium-dependent fluorescence, such that when a neuron is active and signals to other neurons, these proteins inside the cell
become brighter. In this way, light microscopes can measure neural activity at the cellular level. In addition, using so-called optogenetic tools,
they can perturb neural activity by exciting or silencing individual neurons with different colors of light. In zebrafish larvae this can be done
in essentially any neuron by virtue of its transparency and its small size.

Where various microscopy techniques are available, one that is relatively fast and still retains subcellular spatial resolution is *light-sheet
microscopy*. The principle behind this method is to "optically section"
the tissue that is being imaged. This means that at any one time only a
thin plane of the sample is being illuminated, while all other parts of the
imaged volume are kept in the dark.

We and others recently increased the speed of light-sheet microscopy
to such an extent that it can be applied to neuroscience. We built a new

light-sheet microscope that, for the first time, could image activity of almost all neurons in the brain of the larval zebrafish fast enough to track dynamics of neuronal activity (about 80,000 neurons, measured up to a few times per second).

Now, whole-brain interactions between neurons can be investigated; questions can be asked about how groups of neurons "work together" across the entire brain. As a first pass at looking at the neuronal dynamics in the live brain of the larval zebrafish, we measured whole-brain neuronal activity while the fish was doing nothing in particular. It turned out that the brain is very active—a sea of activity is present, including slowly evolving activity, sudden flashes, and, sometimes, large bursts of coordinated activity across the entire brain of the animal (see color plate 2a).

Understanding Complex Whole-Brain Data

The resulting datasets are enormous and generate new challenges: How can one make sense of such a jungle of neural activity? Although networks of neurons communicate in many complex ways, the most basic activity can arise by one neuron simply reflecting the activity level of another, so that if one neuron is highly active, certain others are, too. We searched for such populations of neurons that were active together at some points in time and silent together at others. Such groups of neurons can be pulled out by a relatively simple algorithm: out of the jungle of neural activity arose two populations of neurons in the hindbrain, each showing strongly correlated—or strongly anticorrelated—activity in a set of neurons. These populations had a well-defined anatomical structure, one consisting of six tightly packed, symmetrically arranged clumps of neurons, and the other consisting of two tracts lined by columns of cell bodies (plate 2b, magenta and green, respectively). Within the seemingly confusing tangle of neural activity, these sets of neurons appear to be tightly communicating, with a purpose that hopefully will be discovered in the future.

It is now possible, in principle, to investigate entire sensorimotor transformations and learning processes across the whole brain, from sensory input to behavior. This is especially important because certain brain functions cannot easily be decomposed into constituents and need to be studied at the whole-brain level. Take, for example, a human adjusting its walking pattern to a sudden gust of strong wind to keep it from toppling over. A plausible descriptive mechanism is as follows: circuits controlling movements generate the walking rhythm (which we here call the motor program); the wind hits the body and is reported by sensors in the skin; the vestibular system—the system responsible for the sense of balance—kicks in as it senses an unintended shift in the body's position; both the visual system and the vestibular system signal to the motor system to change its motor program; and higher-order cognitive control makes sure that the change in walking behavior stays in place for the remainder of the journey, until the wind subsides. Since multiple control systems, distributed over many brain areas, are involved in controlling this behavior, it should really be studied at the whole-brain level. The same goes for many other questions; for instance, how do large groups of neurons coordinate to represent an important feature of the visual environment, and how does the entire brain respond to an odor signaling the presence of food and guide the fish to its source? More generally, whole-brain techniques address the very essence of the brain: that all its elements are directly or indirectly in communication with one another.

Future Prospects

Even if we are now in a position to measure a much larger number of variables that would previously have remained unknown—activity throughout the entire brain, the behavior, and all the details of the provided sensory feedback—many challenges remain. Perhaps first and foremost, one must find out what to look for—what problems is the brain actually solving? What aspects of reality (or virtual reality) are important to an animal, and what strategies does it use to approach the challenges it faces? Identifying *what* the brain does—what behaviors in generates, in which circumstances—has to happen alongside studying *how* it does so.

Next, once a series of experiments has generated the appropriate data, what does one do with it? In my lab, we aim to understand how the brain of the larval zebrafish implements behaviorally relevant computations.

Although probably easier than understanding how the human brain works, with its billions of neurons, a comparison to a roughly thirty-cell nervous system that controls the stomach of decapod crustaceans is humbling; after decades of research, this system is still yielding new and surprising insights into the function of nervous systems, and it can be safely said that the thirty-cell network is still not fully understood. This underlines the challenges that lie ahead for understanding a one-hundred-thousand-neuron system. How does one extract principles of brain function from recordings of this network? Computational neuro-scientists, such as our collaborator Jeremy Freeman, are actively working on methods for analyzing the ever-larger datasets that neuroscience generates. The more data we have about the nervous system, the tighter we can constrain models of it. In a real sense, whole-brain imaging with behavior, and computational neuroscience, are perfect partners (see chapters by Shenoy and Freeman, this volume).

On a more technical note, our current method for measuring activity in almost all neurons is relatively slow compared to the millisecond timescale at which neurons communicate: each neuron is observed roughly once per second. To look at how neurons communicate at millisecond timescales, speed increases will be necessary by refining or developing new microscopy techniques used for volumetric imaging. Similar improvements in the genetically encoded sensors of neural activity are necessary.

Finally, measurements of neural activity are, of course, not enough for understanding the system. Functional observations have to be understood mechanistically; given a neuron's behavior in the midst of the sea of activity of other neurons, can we understand one in terms of the others? To constrain this problem and to really understand how neurons communicate, we must know the wiring properties between the neurons. Genetic techniques, and methods based on electron microscopy, will complete the picture by providing such anatomical information; work on generating such data has already begun. Finally, *perturbing* the dynamical system rather than just observing it is another necessary ingredient for understanding it. Can we build conceptual and predictive models for how this brain works, then make a prediction of what it would do when a set of neurons suddenly falls still? These types of hypotheses need to be tested, and optogenetic tools such as channelrhodopsin and halorhodopsin, capable of exciting and silencing neurons when hit by different colors of light, allow for this and have already been used in a series of elegant studies that explore the causal role of genetically or anatomically defined neuronal populations.

The brain is an organ that evolved within the dynamical environment that consists of the body and the external world, and is the "first responder" within that system (it responds on a timescale of a hundred milliseconds; other organs are much slower). The meaning of "understanding brain function" depends in part on the researcher asking the question, but, most likely, the path to a holistic understanding requires studying it on all the different scales; we need to understand molecular mechanisms and single-neuron dynamics to construct a picture of how network function arises from these building blocks. Conversely, to understand how these collaborate and form the whole that is greater than the parts one needs theories of *global* brain function. Hopefully, holistic approaches such as those described above will help us construct just that.

Further Reading

Ahrens, M. B., J. M. Li, , M. B. Orger, D. N.Robson, A. F. Schier, F. Engert, and R. Portugues. 2012. "Brain-wide neuronal dynamics during motor adaptation in zebrafish." *Nature* 485 (7399): 471–77. doi:10.1038/nature11057.

Ahrens, M. B., M. B. Orger, D. N. Robson, J. M. Li, and P. J. Keller. 2013. "Whole-Brain Functional Imaging at Cellular Resolution Using Light-Sheet Microscopy." *Nature Methods* 10 (5): 413–20. doi:10.1038/nmeth.2434.

Chiel, H. J., and R. D. Beer. 1997. "The brain has a body: Adaptive behavior emerges from interactions of nervous system, body and environment." *Trends in Neurosciences* 20 (12): 553–57.

Harvey, C. D., F. Collman, D. A. Dombeck, and D. W. Tank. 2009. "Intracellular dynamics of hippocampal place cells during virtual navigation." *Nature* 461 (7266): 941–46. doi:10.1038/nature08499.

Seelig, J. D., M. E. Chiappe, G. K. Lott, A. Dutta, J. E. Osborne,M. B. Reiser, and V. Jayaraman. 2010. "Two-photon calcium imaging from head-fixed Drosophila during optomotor walking behavior. *Nature Methods* 7 (7): 535–40: doi:10.1038/nmeth.1468.

PROJECT *MINDSCOPE*

Christof Koch

With Clay Reid, Hongkui Zeng, Stefan Mihalas, Mike Hawrylycz, John Philips, Chinh Dang, and Allan Jones

The human brain, with its eighty-six billion nerve cells, is the most complex piece of organized matter in the known universe. It is the organ responsible for behavior, memory, and perception, including that most mysterious of all phenomena, consciousness. Neuroscience, the discipline that seeks to understand the principles underlying the brain's operation, has over the past century and a half of its history uncovered its constitutive elements—membrane channels, synapses, and nerve cells. Yet their stunning heterogeneity, sheer numbers, and the breathtaking diversity in which they are assembled has resisted reductionist understanding of anything but minute aspects of its behavior. Furthermore, given the interbraided nature of the nervous system—with many neurons receiving input from literally thousands of other neurons and making output with thousands—an inexhaustible multiplicity of factors influence any one neuronal action. Yet understand it we must! Not only because the relationship between objective events in the brain and subjective phenomena in the mind remains one of the deepest scientific puzzles but also because of the intolerable toll that nervous system pathologies and injuries take on individuals, their families, and society at large.

Given the difficulty of the task and the brief span of our life, let us focus on a more circumscribed problem: that of understanding how information is represented and transformed in the neocortex, the proverbial gray matter of the brain. The neocortex is a layered structure whose thickness varies by a factor of about two while its surface area varies by fifty thousand between the smoky shrew and the blue whale. A unique hallmark of mammals, the neocortex is a highly versatile, scalable computational tissue that excels at real-time sensory processing across modalities, making and storing associations, and planning and producing complex motor patterns, including speech. The neocortex consists of smaller modular units, columnar circuits that reach across the width of the cortex, repeated iteratively within any one cortical area. These modules vary considerably in connectivity and properties among regions. The computational function of a neocortical column—filtering the input, detecting features, context-dependent amplification, look-up table, line attractor, predictive coder, and so on—remains unclear, with some controversy whether there is, indeed, a single canonical function performed by any and all neocortical columns. Yet neocortical architecture and genomic expression pattern is remarkably constant across species and regions—albeit with many exceptions. And unlike worms and flies with their high degree of stereotypy, in which genetically determined neural circuits mediate innate behaviors, mammalian neocortical circuits are shaped by the experiences of their ancestors, in the form of genetic specialization within cortical regions, as well as by personal experiences in the form of synaptic learning, and exploit more general-purpose, flexible population coding principles that are highly sensitive to context. In that sense, the cortical column may be the closest that nature has come to evolving a universal Turing machine, a machine whose settings are adapted by a combination of genomic and learned (synaptic) mechanisms to the particular statistics of its input, be it visual, olfactory, linguistic, or otherwise.

Of Men and Mice

A deep understanding of the cortex necessitates querying the relevant microvariables, in particular spiking neurons, by recording the occurrences and timing of action potentials. Active neurons rapidly assemble and disassemble into far-flung coalitions that can be tracked from the sensory periphery to motor structures. Mapping, observing, and intervening in such widespread but highly specific cellular activity can best be accomplished by shifting emphasis from humans (in which many techniques are ethically precluded) to an evolutionary related model organism that allows comprehensive measurement and intervention, that of the mouse, *Mus musculus*. The two species, whose last common ancestor lived around seventy-five million years ago, share much of their genome. Indeed, 99 percent of mouse genes have a direct counterpart in the human genome, with an average 85 percent similarity level at the nucleotide level. Data from the Allen Mouse and Human Brain Atlases indicate that there are a myriad of differences between the cellular-level expression of genes between mice and humans (just as there are between any two species). The biggest salient differences at the level of the cortex can be found between visual and somatosensory cortices, reflecting the importance of foveal vision for humans and whiskers for mice.

Beyond these, there are two very obvious differences between the human and the mouse brain—accessibility and size. First, for obvious ethical reasons, the living human brain can only be probed at the required cellular level under rare conditions, primarily for neurosurgery. Conversely, with appropriate care for the well-being of the animal, the smooth neocortex of the mouse is fully accessible to electrophysiological and optical imaging techniques. Furthermore, experiments with appropriately modified viruses to stain, mark, turn on, or turn off molecularly identified subpopulations of neurons permit unprecedented control of mouse brain circuitry. This cannot be emphasized enough. The exploding use of opto- and pharmacogenetics methods that delicately, transiently, reversibly, and invasively control defined events in defined cell types at defined times constitute a suite of interventionist tools that allows neuroscience to move from correlation to causation, from observing that this circuit is activated whenever the subject is contemplating a decision to inferring that this circuit is necessary for decision making. Second, the human brain is more than three orders of magnitude larger than the mouse brain—1.4 kg weight versus 0.4 g; a 1-liter volume versus a sugar cube; eighty-six billion nerve cells versus seventy-one million for the entire brain and sixteen billion versus fourteen million nerve cells for the neocortex.

Project *MindScope*

The overall similarity of the mouse and the human brains, and the much smaller size of the former, makes it feasible to mount a large-scale, comprehensive, and focused effort to map out the detailed circuitry at the cell-type specific level and to record, visualize, perturb, and model the spiking activity of a significant fraction of all cortico-thalamic neurons underlying a few, archetypical behaviors in mice. In this manner, understanding how sensory information is coded, transformed, and acted upon at the timescale of a one or a few perception-action cycles (< 2 sec.) can be achieved by a sustained, focused, and high-throughput effort that encompasses five, tightly interwoven, research strands. We call this effort project *MindScope*.

MindScope is part of a ten-year, high-throughput, milestone-driven effort announced in March of 2012, an endeavor involving several hundred scientists, engineers, and technicians. Philanthropist Paul G. Allen, who founded the institute in 2003, pledged US$300 million for the first four years of this ambitious plan. He also committed to the construction of a new 270,000-square-foot building in Seattle, to be occupied in late 2015.

MindScope focuses on the visual system and visuo-motor behaviors in the young adult laboratory mouse. Institute scientists seek to understand the computations that lead from photons to behavior by observing and modeling the physiological transformations of signals in the cortico-thalamic visual system (see color plate 3). Participants want to catalog and characterize the cellular building blocks of the cortico-thalamic complex, their dynamics, and the cell-type specific and individual connectomes. Scientists want to know what the animal sees and how it thinks in a quantitative manner. This requires the tight integration of results across distinct methods and scientific disciplines—classical and molecular neuroanatomy, electro- and optical-physiology in behaving animals, computer modeling of the cellular populations and their dynamics, and theoretical considerations concerning the design and operation of the overall system.

Mapping Out Cell Types and Their Connectivities

To achieve these ambitious goals, we rely on a set of genetically engineered mice that have one or a few specific neuronal cell types marked in each mouse. Taking advantage of the unique gene expression patterns of each neuronal cell type, we create transgenic mouse lines in which the promoters of specific genes are used to express a master control gene, most often the Cre recombinase. Such *Cre driver lines* (see glossary) are used to effectively induce controlled mutations in specific cell types or at a specific point in time. They can also be combined with a variety of in-house reporter mouse lines or engineered viral vectors to control the expression of various effector genes to label and manipulate neurons. Such transgenic mice are powerful genetic scalpels that enable us to dissect circuit components. We have constructed more than forty such Cre driver lines that comprehensively cover excitatory and inhibitory cell types in the cortico-thalamic circuit continue to generate more refined ones as our knowledge about different cell types grows, and have made them publicly available through the Jackson Laboratory.

On the effector side, we incorporate state-of-the-art molecular tools for monitoring and manipulating circuits. These include various fluorescent proteins (such as GFP) for visualizing neuronal morphology and connectivity, genetically encoded calcium indicators such as GCaMP6 (see chapter by Ahrens, this volume) for reporting neuronal activities, and light-driven opsins such as channelrhodopsin for altering neuronal activities. That is, once we have identified a molecular zip code for any one set of neurons, we can label this set and turn it on or off for anywhere between milliseconds to hours. These genetic tools provide the foundation for our experimental investigations.

Several years ago, we embarked on a large-scale effort to develop a regional and cell-type specific three-dimensional connectivity map (projectome). This Allen Mouse Brain Connectivity Atlas (see chapter by Hawrylycz, as well as www.brain-map.org) uses a genetic tracing approach and a high-throughput, serial two-photon tomography system to image the GFP-labeled axons throughout the entire brain in a pipelined manner in thousands of mice (see below). We couple high-speed two-photon microscopy with automated vibratome sectioning of an entire mouse brain. High-resolution coronal images are sampled every 100 μm, resulting in a 0.75 TB dataset per brain. About 2,000 such datasets have already been generated, all registered into a common 3D reference space with high spatial fidelity. This allows quantitative analyses of the entire dataset, parsed into half a million 100 x 100 x 100 μm^3 pixels, and mapped onto 295 anatomically defined regions based on the Allen Reference Atlas ontology tiling all of brain space. The next step is an even more focused exploration of the visual cortico-thalamic circuitry and its associated cell-type specific connectivity using a true 3-D atlas with 10 μm isotropic resolution.

We will comprehensively characterize the physiological, anatomical, and transcriptional properties of cell types in the visual cortico-thalamic system (we estimate < 100 such types). Our goal is to derive a comprehensive taxonomy of cell types for this circuit. Biological heterogeneity at cell-type and single-cell levels may underlie the functional diversity, flexibility, and plasticity of the neural circuits, as well as the genetic underpinning and environmental influence unique to each individual animal.

We plan a three-pronged attack at the single cell level. Firstly, we will record the full spectrum of biophysical and intrinsic firing properties of cell types both in slices and in the living brain under a standard electrophysiological protocol from patch-clamp recordings taken at the cell body. Furthermore, we will use generalized leaky integrator models to accurately replicate the observed subthreshold voltage and spike timing behavior using computer simulations. Secondly, we will image and reconstruct full morphologies of thousands of neurons to characterize their dendritic tree and their proximal and distal axonal arborizations throughout the brain. Thirdly, we aim to classify cell types by partial (qPCR) or near-full (RNA-seq) read-out of the mRNA species transcribed in thousands of individual cells. Finally, we will investigate the detailed physiological properties, in particular the short-term changes in synaptic strength, of synaptic connections between partner neurons of the same or of different types. This vast amount of data will feed into our large-scale modeling efforts to generate conceptual and realistic circuit models and theories about circuit computation (see below). As in our past efforts, these data will be freely and publicly available in an online, curated database of cell types.

While the morphology and biophysical properties of cell types gives an outline of the connections between cortical neurons, highly interconnected local cortical circuits have far more structure than the statistics of pair-wise measurements would predict. In addition to statistical rules of regional and cell-type specific connectivity, higher-order connectivity rules among groups of cells relate to functional specificity. We are therefore planning further anatomical studies of fine-scale connectomics in the visual cortex: the study of synaptic networks (w_{ij}) between individual neurons (i and j) in cortical circuits. These studies will include large-scale reconstructions of cortical circuits with serial-section electron microscopy as well as more targeted experiments using rabies virus for trans-synaptic labeling. Because specific individual connections in cortical networks are likely to be related to specific physiological responses, these anatomical experiments will be carried out in animals in which single neurons had previously been studied physiologically during behavior.

Recording the Activity of a Large Ensemble of Neurons

The ultimate goal of *MindScope* is to understand the computations performed by the cortex. We chose the visual system as our entry into the mind of the mouse. Although its visual acuity is about 50x lower than that of humans with their fovea, it retains most of the anatomical and physiological features that make vision the best-studied sensory modality. While much is known about whisker- and olfactory-triggered behavior in mice, the range of visuo-motor behaviors mice are capable of remains to be explored.

Visual information originates in photoreceptors. In the common laboratory mouse that we use (C57BL/6J), there are about six to seven million rods and 180,000 cones in each eye. The resultant analogue information percolates through the retina and generates action potentials in about 50,000 ganglion cells. These come in about twenty flavors with distinct morphologies, response patterns, and molecular signatures, each one of which tiles visual space. Many of these project to the 18,000 neurons of the lateral geniculate nucleus (LGN), part of the visual thalamus. The axons of LGN cells in turn connect with some of the 360,000 neurons of the primary visual cortex (V1). Detailed anatomical tracing has revealed its rather elaborated hierarchical structure with about a dozen visual associational cortical areas surrounding V1 (compared to at least three dozen in primates). The visual information is further processed while passing through V1 and associational cortical areas, making up a network of networks (see color plate 3).

We are planning a series of increasingly complex visuo-motor behaviors, centered around visual invariant object recognition, to record how these behaviors influence receptive field and other properties in V1 and how the cortical circuitry in turn shapes behavior. We can train mice to make a discrimination between two simple stimuli (for example, a left- versus a right-tilted grating) while they run and assess the effects of disrupting neural activity in specific cell types, cortical layers, or cortical areas while monitoring the activity of large ensembles of neurons.

Classically, the spiking activity of neurons has been studied with electrophysiology. Multielectrode recordings, previously achieved with handcrafted electrical probes, have entered the silicon age. Commercially available probes with up to sixty-four sites per shank are currently available, leaving room for dramatic improvements. We are engaged in a consortium with other institutions to create high site-count silicon probes, with integrated circuitry to amplify and multiplex the signals from 512 sites per single shank. Active circuitry at the base of the shank will amplify, filter, and multiplex these signals such that only a handful of wires will have to be connected off chip. The goal in the coming years is to record from a large fraction of the cortical neurons making up a minicolumn. The assignment of these neurons to specific cell types has now becoming possible with the Cre-mediated expression of channelrhodopsin, which makes specific cell types directly excitable by light for identification.

Functional imaging of neurons in the behaving mouse has reached the point that recording the activity of hundreds of individual, genetically targeted neurons has become routine. Two-photon microscopy visualizes fluorescently labeled neurons with sufficient resolution to identify individual spines. Combined with genetically encoded calcium indicators such as GCaMP6, which provide signals that are dominated by action potentials in the soma, it is possible to resolve bursts of spikes in hundreds of neurons at video rates (of course, this is considerably slower than the submillisecond resolution of electrophysiological recordings). Because all of the visual cortical areas lie directly below the surface of the skull, we can target each of these areas for physiological imaging, sometimes in the same experiment. Recently it has even become possible to record the spiking activity of individual axons and presynaptic boutons, so that the signals provided by projection neurons can be recorded in the target area. Therefore, in addition to the interareal anatomical connectivity map (projectome) described above, we plan to delineate a functional projectome: a comprehensive map of the physiological signals sent between visual areas from identified cell types during different states of the animal (including deep sleep, running, and one or more visuo-motor tasks involving spatial attention and invariant object recognition).

Structured Science: Bringing Quantitative and High-Throughput Tools from Industry to Neuroscience

We seek to greatly speed up the acquisition of relevant biological data by linking together multiple platforms, from surgery to electron-microscopic reconstruction and annotation of circuits, from behavioral suites to two-photon calcium imaging, into high throughput, standardized pipelines using quality control (QC), standard operating procedures (SOP), milestones, and other tools that are de rigueur in the biotechnology industry. Core to every one of our atlasing projects at the Allen Institute has been a clear definition of our target product and a methodical mapping of milestones and deliverables through a detailed project-planning and management process. We combine the appropriate multidisciplinary scientific and technical teams, including biologists, modelers, data analysts, and engineers, to industrialize processes and execute on delivering the product on time and within budget.

Each module is linked into a large-scale data generation pipeline where standard operating procedures (SOPs), supply chain management, and careful quality control measures are all employed in a high-throughput setting. Once generated, data enters our Informatics Data Pipeline where it goes through multiple rounds of processing and quality control on its way to becoming part of one of our freely available online products. Depending on the nature of the work going through our Structured Science laboratories, any one module might be supporting multiple data generation pipelines or directly supporting exploratory work by one of our research scientists.

Our newest online data product, the Allen Mouse Brain Connectivity Atlas (www.brain-map.org), is an excellent example of our large-scale pipeline approach. We industrialize transgenic mouse generation and characterization, stereotaxic rAAV tracer injections, and serial two-photon tomography, and link these data generation modules to a robust informatics data pipeline to generate the main datasets for the atlas product (figure 1). To meet the goals of this project, we scaled up our ability to generate, characterize, and breed large numbers of Cre driver mouse lines and to coordinate these activities with a dedicated core of surgeons that carried out precise rAAV tracer injections on a daily basis for several years. This pipeline provides thousands of injected mice brains for serial two-photon tomography using six TissueCyte 1000 systems. Each system sections and images one brain per day (sampled at 0.35 μm in x-y, much less in z), six days a week, in an industrialized setting where we fully monitor the ~20 hour runs and have on call staff ready to support any run problems outside of normal business hours. The output from the first three data generation modules is raw two-photon image data from a matrix of injections in wild-type and Cre driver line mice. These data are then ready for multiple stages of processing and quality control as part of the complete pipeline.

This includes automated data processing, search, visualization, and analysis of a large and complex dataset. Developed as an enterprise system for scalability, it consists of a number of algorithmic modules that are integrated into an internally developed laboratory information management system (LIMS) and job scheduling and submission backbone. These components form a fully automated pipeline, capable of processing one petabyte (10^{15} bytes) of imaging data per year.

The final product is delivered via a web application at www.brain-map .org. It allows users to search for projections of one or more structure(s) to other structure(s), search similar connectivity patterns and for virtual retrograde connections, view and download the original primary data in high resolution, view data in both 2D and 3D along with the Allen Reference Atlas to provide anatomical context, view and analyze one or more datasets and download all computed values to perform large-scale data mining and analysis.

We plan to apply similar project management techniques to the generation of high-quality data products for understanding the structure and function of mouse cortical neurons and cortico-thalamic circuits under both *in vitro* and *in vivo* conditions.

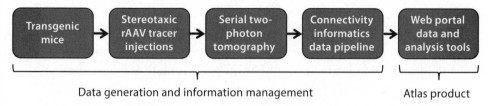

Figure 1. Allen Mouse Brain Connectivity Atlas data generation and information management pipeline.

Modeling the Brain by Integrating Diverse Data and Theory

The sheer complexity of the nervous system makes development of quantitative biophysical and more abstract point neuron and network models critical to understand their function. There are two sets of observables that relate to the function of the organism and which brain models will need to reproduce: neuronal activity (subthreshold membrane potential and spiking behavior) and the animal's choice behavior. Even though we initially plan to use a fairly naturalistic foraging behavior, running on a wheel, this represents only a small subset of possible behaviors the animal is capable of. We seek to construct models that have a high likelihood of generalizing to as of yet unobserved behavior. As such, we restrict ourselves to mechanistic models that reproduce characteristics of neuronal activity. The level of detail over which the model aims to reproduce the neuronal activity, population statistics of identified cell types, and responses of individually characterized neurons is determined by the levels of resolution at which we plan to quantify the mouse connectome, both cell type–specific connections, $w_{\alpha\beta}$, as well as individual connections, w_{ij}. Since the vast majority of the information passed between cortical neurons are spikes, we will build a series of visual models linking structure to spikes.

We plan to parameterize a series of synaptic, neuronal, and small circuit models that reproduce neuronal input-output relations. Models for LGN and V1 neurons will be constrained by correlated slice electrophysiology, morphology, and transcriptomic data. For synapses, the focus is on the correct parameterization of short-term plasticity between the multitudes of cell types. We will initially use generalized leaky integrate-and-fire units to model neurons, as known convex optimization methods allow us to find a unique set of parameters that best fit the measured subthreshold voltage and spiking behavior. Biophysical models have an intrinsic beauty, as they incorporate a wealth of information known about the system on top of the point neuron models: morphology, voltage-dependent currents and synapses and their locations, and rely exclusively on their physical description. However, they are very difficult to parameterize and parameter searches is one of the most computer-intensive part of modeling. An open question is whether there are canonical circuits that occur often enough to be analyzed as modules: cortical interneurons-pyramids microcircuits, cortico-thalamic loops, and so on. As it is true for all of biology, discovering modularity is the key principle that allows us to avoid what has been referred to as the *complexity brake*. Since most of the components are parameterized from slice data, we have to incorporate both modulatory and direct synaptic input from other cortical and thalamic regions.

A large number of the brain regions connect to and influence activity in the visual areas, some set the level of vigilance, while some provide auditory, somatosensory, or motor signals. To quantify this background state, we developed a population statistic method that reproduces the subthreshold and the spiking activity of a population of simple neuronal point models under conditions of homogeneous synaptic input. It can be applied to very large networks, up to the entire brain to mimic the sleeping brain or "resting state" activity. The essential parameters for such a model are the cell-type connectome and dynamical properties of neurons and modules.

For the system-level models, we plan to build network models that incorporate all the processing levels and modules we characterize and to use modeling description languages and other simulation tools that allow inclusion of hybrid modules (mix and match); for example, visual areas outside of the primary visual cortical area (V1) described by population statistics, V1 by spiking populations, and a limited number of neurons by biophysical details, aimed at reproducing *in vivo* physiology. This multi-resolution approach limits the computation time spent on regions that have modest effects on simulations outputs (as characterized by sensitivity analysis) as well to ease the understanding of the simulation results.

It will be critical to characterize system models in terms of their spatiotemporal receptive fields; that is, the region in visual space from which a precisely timed series of visual stimuli can evoke a response (as in a direction-selective visual neuron that best responds if a bar is moved in one direction). Subsequent analysis steps involve behavioral state dependence and context integration in the processing of a natural visual scene the mouse is likely to be sensitive to. We also plan to reconstruct the visual input based on the observed neuronal activity and models of what the neurons code for in an image (so-called mind-reading techniques using Bayesian methods). These reconstructions will allow us to test specific coding models in a quantitative manner: for example, is the quality of reconstruction a function of how salient these features are? While our emphasis is on models of stimulus encoding, decoding stands in a dialectic relationship to encoding—for the better our understanding of how neurons encode stimuli, the better decoding will be.

A final step in constructing models is to link them with the observed animal choices in a task. A hybrid model containing components at very different levels that is mapped to behavior is very useful in predicting which components and which details do contribute to observable outcomes using sensitivity analysis techniques. This analysis is crucial for making predictions of the consequences of opto-genetic interventions on network activity.

Yet even if we could record from every single spike from every neuron in the brain and could simulate every such spike in a gargantuan biophysical simulation, we would still not understand the deep principles underlying its processing (see also chapter by Freeman, this volume). Answering questions such as, What is the function of feedback pathways to the thalamus? Is the brain organized in a series of linear-nonlinear feedforward processing stages modulated by feedback (as in the most popular model of primate vision)? How are objects represented? How can these representations be manipulated and learned? To what extent does the cortex perform predictive coding or Bayesian inference to adjust its settings? and so forth, requires a concerted theoretical effort that works hand in glove with the experimental and the modeling efforts. Such theoretical considerations will complement our modeling efforts.

As modelers, theoreticians, anatomists, and physiologists are all sited together and work on the same cortico-thalamic system, both modeling and theory will guide our future experimental investigation, establishing a tight virtuous loop between what can be measured, what should be modeled, and what can be understood.

Toward Large-Scale, Open-Source Science

Neuroscience is a splintered field, with worldwide about ten thousand independent laboratories pursuing distinct questions, across spatiotemporal scales ranging from nm to cm and from μsec to years, with a dizzying variety of tools. In a 2012 commentary in *Nature*, Koch and Reid wrote about the challenges of studying the brain under these conditions in the era of Big Science. Because students must write first-author papers to graduate, and faculty must publish in high-impact and hypercompetitive journals to obtain grant support and tenure, the modern academic research enterprise encourages maximal independence among experiments and groups. Indeed, when attending the annual Society for Neuroscience meeting one is struck by the speed with which its sixty thousand or more practicing neuroscientists are heading away from each other in all possible orthogonal directions in a sociological form of Big Bang. While this is necessary in the romantic, exploratory phase of any science, a more systematic and thorough exploration of canonical circuits and behaviors is called for as neuroscience enters a more mature phase.

This orthogonality among groups has prevented the emergence of common standards and a handful of canonical, large-scale projects. For example, there is no universally accepted definition for identifying action potentials in noisy electrical recordings from different neuronal tissues. Instead, dozens of distinct spike detection and classification algorithms are in use. In order to gain a competitive edge and for lack of funding to manage and curate online repositories of data, hard-gained information is amassed and rarely made accessible online. Molecular compounds and transgenic animals are only shared after the initial papers describing them have been published in print. All of this has made comparison across laboratories difficult, replication of specific experiments arduous, and has significantly slowed progress.

MindScope is also an experiment in the sociology of neuroscience, heralding the arrival of large-scale science in a field populated by small groups. It will require rewarding the team for the collective effort rather than rewarding a few lead investigators. Integrating distinct scientific methods originating with an individualistic academic environment with a more team-driven corporate approach constitutes the true challenge of *MindScope*. By assembling a large team of specialists focused on a common set of goals, techniques, and standards, *MindScope* will achieve much more than any one specialist can on their own. For only then can the massive and disparate anatomical, imaging, and physiological data be synthesized into a mathematical and predictive framework of how all elements fit together and act as a whole to give rise to intelligence and consciousness. We envision a day, not far in the future, where a small number of highly sophisticated, equipped, and staffed brain observatories will complement the academic landscape.

While *MindScope* seems daunting, other scientific fields have successfully carried out much more massive undertakings, such as the construction of high-energy particle accelerators, telescopes, or the human genome effort. These involve hundreds to thousands of scientists, engineers, and technologists and operate over a timescale of decades with commensurate budgets funded by national governments, foundations, and private donors. In a manner comparable to how physical scientists build instruments to gaze at distant events at the edge of the universe, brain scientists must build observatories to peer at proximal events inside the skull that give rise to the mind that wonders and peers.

Acknowledgment. None of this would be possible without the unprecedented generosity and long-term vision of our institute founders and benefactors, Jody and Paul G. Allen. This generosity and vision enables us to do things that have never been done before and to significantly speed up the arrival of the day when we will understand the human brain and the many pathologies it suffers from.

THE CONNECTOME AS A DNA SEQUENCING PROBLEM

Anthony Zador

Each year more than thirty thousand neuroscientists gather to share what they have discovered, enough to fill thousands of scientific papers. The rate of progress is staggering. Yet we still don't really understand how the brain works.

Why? I would argue that the reason we don't understand how the brain works is that we are missing crucial information. Although we know a great deal about molecules and single neurons, and also about the gross organization of brain areas, our knowledge is scarce between these two extremes, at the level of neural circuits. For this a vital prerequisite is knowing the wiring diagram. The good news is that because of recent advances in technology, it may soon be possible to obtain the wiring diagram, or "connectome," of the brain at single neuron resolution.

Proof that we are not there yet—that we still haven't "solved" the brain—comes from the fact that we are still apparently quite far from being able to build one. If we really understood the principles behind thought, we could build a machine capable of humanlike thinking.

But so far we can't. At the dawn of the computer age over half a century ago, expectations were high that computers would soon perform many of the same cognitive functions that humans do. Herbert Simon, one of the fathers of artificial intelligence (AI), predicted in 1965 that "machines will be capable, within twenty years, of doing any work a man can do." Of course, these predictions turned out to be wildly off the mark.

It soon became clear that some cognitive functions were harder to train computers to perform than others. The surprise was that tasks that were easy for a person often turned out to be hard for a machine, and vice versa. Many apparently simple tasks, on which toddlers make considerable progress in their first two or three years—drinking from a cup, tumbling with a friendly dog, or identifying the villain in an animated fairy tale—are very challenging for machines. Computers today play better chess than any human world champion, but because of the primitive state of machine vision and allied fields we still don't have domestic robots loading our dishwashers.

Why is that? Is there something special about biological computation that renders it so superior in some domains? Or could such functions be replicated with classical computer architecture? At the risk of greatly oversimplifying a very complex history, one might say that different answers to this question have historically led to two very different approaches. On the one hand, those who held that there was nothing special about biological wetware went on to pursue what are now thought of as classical AI approaches. On the other hand, some researchers believed that the style of brain computation is what makes biological systems superior, and that only by building computational engines on those same principles would we match the capacity of real organisms. This latter view led eventually to the field of connectionism—neural networks and machine learning.

Although the roots of neural networks can be traced back to the 1950s and even deeper, one might conveniently date the modern Renaissance of the field to the publication of Rumelhard and McClelland's PDP books in 1986. The parallel distributed processing (PDP) manifesto proposed that the key features of brain-like computation were that it was parallel and distributed. Many simple summation nodes ("neurons") replaced the single central processing unit (CPU) of computers. The computation was stored in the connection matrix, and programming was replaced by learning algorithms such as Paul Werbos's backpropagation. The PDP approach promised to solve problems that classic AI could not.

Although neural network and machine learning have proven to be very powerful at performing certain kinds of tasks, but they have not bridged the gap between biological and artificial intelligence, except in very narrow domains, such as optical character recognition. What is missing? One possibility is that even neural networks are not "biological" enough. For example, in my PhD thesis I explored the possibility that endowing the simple summation nodes of neural networks with greater complexity, such as that provided by the elaborate dendritic trees of real neurons, would qualitatively enhance the power of these networks to compute. But the advantages turn out to be quantitative only; adding that particular sort of biological fidelity scarcely allowed us to span the biological-computational gap as we had hoped. An alternative and more popular idea was and is that we need to develop more sophisticated learning *algorithms*. Indeed, what began as one of the leading conferences in the neural network field, Neural Information Processing Systems (NIPS), quickly evolved into a conference focused almost exclusively on machine learning. But thus far decades of work in machine learning have not sufficed to decrypt the brain.

At this point we must take seriously the possibility that neither more biologically realistic networks nor better learning algorithms will solve the problem. Rather, biological organisms may be very effective at performing certain computations because they have evolved a highly specialized set of finely optimized algorithms for solving them, a "bag of tricks." These "tricks" may deal with the many special cases and exceptions needed to render the algorithms effective over a wide range of real-world scenarios. This view is anathema to theorists in search of a small set of unifying principles to explain biological computation. But perhaps it is sensible when one considers that organisms have been subjected to several hundred million years of selective pressure to evolve an effective bag of tricks. The brain might be what Gary Marcus has called a "kluge," a clumsy and inelegant solution that gets the job done, without necessarily being beautiful.

The bag-of-tricks hypothesis does not imply that good, general-purpose algorithms are unnecessary, only that they are not sufficient. Even the best bag of tricks in the hands of an amateur magician doesn't make for a good magic show. In the same way, the kernel of Google's success as a search engine lay in the PageRank algorithm (which ranks pages according to the number and quality of links pointing to that page), but Google's current preeminence in search stems from careful tweaking of (according to Google) over two hundred other "clues," or tricks (http://bit.ly/1fTz2C6), such as the freshness of the page and the user's geographical location. The bag-of-tricks hypothesis raises the possibility that biological intelligence may represent the distillation of such a vast "training set"—the life-and-death experiences of our myriad ancestors over hundreds of millions of years of evolution—that even the most sophisticated learning algorithms may fail to discover them, simply because their training sets are far too small.

I have reluctantly embraced the bag-of-tricks hypothesis as the explanation for why biological intelligence continues to outshine artificial intelligence—reluctantly, because this view implies that there is no grand insight that will reveal to us how the brain works, and catalyze the creation of intelligent machines. Instead, the bag-of-tricks model suggests that if we are to build machines that perform well on certain classes of real-world problems, we must either dissect the tricks biological intelligence uses or invent our own. Connectomes, at the single neuron level, may allow us to provide the information we need to reverse engineer the brain.

The good news is that the tools and techniques needed are almost within our grasp.

The natural circuit to study for insight into biological intelligence is the neocortex, the structure widely credited with endowing us with our intelligence. The cortex is uniquely mammalian and achieves greatest elaboration in primates, especially humans. The basic structure of the cortex is largely preserved across mammals, so that a bit of cortex from a rodent does not appear very different from the corresponding bit in a monkey. Within an organism, cortical structure is fairly uniform, so that a bit of cortex involved in processing sound is not that different from another bit involved in processing touch.

These and other considerations suggest that the cortex is a modular structure. It appears that protomammals may have evolved the cortex to solve the very challenging problem of scaling of neural circuits. Presumably having a bigger brain endows an organism with an evolutionary advantage in terms of greater behavioral flexibility, but from an evolutionary perspective it is not necessarily straightforward to incorporate new neurons into a circuit. Circuit architectures that work for small circuits do not necessarily work for large circuits. For example, all-to-all connectivity scales quadratically with the number of neurons (all-to-all connectivity in a 10 neuron circuit requires only 100 connections, whereas in a million neuron circuit a thousand billion connections would be needed), so quickly becomes impractical as the number of neurons grows. Furthermore, modularity helps solve the development problem of wiring up a brain. In some organisms, such as the worm *C. elegans*, the entire neural circuit (consisting of 302 neurons and about 7,000 synapses) is specified precisely by the genome, but specifying each connection in the genome quickly becomes impractical as the number of bits needed to specify all the connections exceeds the number of bits in the genome.

The basic cortical module is often taken to be the same as the cortical column—a vertically organized collection of cortical neurons that respond similarly to, for example, sensory input. However, understanding the cortical module requires not only that we understand the local circuitry within a column but also the inputs and outputs of the module. A given cortical region is intimately associated with other cortical regions, as well as with other structures such as the thalamus and striatum.

The modularity of cortical architecture gives us hope that we will understand biological intelligence. It suggests that our goal should be to understand the basic organization of the cortical module, how it is wired together with other modules, and how modules from different brain areas are specialized to perform specific functions. The circuit architecture common to most or all cortical modules may reflect the basic shared structure of cortical algorithms, whereas circuit motifs found only in specialized cortices may reflect special tricks needed to process specific kinds of information. Understanding these basic principles of cortical organization will lay the foundation for decoding the bag of tricks encoded in the wiring of neural circuits. Note that in this view, there need not be anything fundamentally special about the cortex. Rather, understanding it is merely a prerequisite for decoding the bag of tricks, in the same way that learning FORTRAN is a prerequisite for learning many algorithms useful in numerical analysis—algorithms that can then readily be reimplemented in C or any other language. Once we understand the basic principles of cortical computation, there is no reason to think we could not reimplement them in an artificial silicon brain.

To reverse engineer biological intelligence we must understand how specific neural circuits solve well-defined problems. The obvious model organism in which to conduct such studies today is the mouse. Mice are genetically accessible, which allows us to bring to bear the full armamentarium of modern molecular biology. Rodents can be trained to perform sophisticated sensorimotor decision tasks similar to those used in nonhuman primates. Moreover, it is now possible to monitor the activity of hundreds or thousands of neurons simultaneously, using calcium imaging (see chapter by Ahrens, this volume). Other approaches, such as the DNA ticker tape proposed by Konrad Körding (Northwestern) and George Church (see his chapter herein), raise the possibility that we may eventually be able to record from even more. Thus it is possible to record and manipulate neural activity in animals performing well-defined behaviors that require specific computations.

What has lagged behind is the technology for unraveling the detailed wiring of the circuit, the "connectome." Current approaches to determining the connectome are based almost exclusively on microscopy. Unfortunately, microscopy is poorly suited to the study of neural connectivity because brains are macroscopic structures, whereas unambiguous determination of a synaptic connection requires electron microscopy (EM). So far, the complete connectome has been established for the worm *C. elegans*. However, determining even this simple connectome (302 neurons connected by ~7,000 synapses) was a heroic task, requiring over fifty person-years of labor.

There are two main challenges to using EM to reconstruct the connectome. Reconstruction based on EM requires imaging many very thin (~10 nm) 2D tissue sections and then aligning successive sections to infer the 3D structures from which they were derived. The first challenge is that acquiring the data is very difficult. Traditional EM methods have neither the requisite throughput nor accuracy—even a few lost sections can severely compromise the reconstruction, necessitating extremely reliable sectioning methods. The second challenge is analysis. Inferring the 3D structure from 2D sections requires matching the corresponding neuronal structure in each successive image. Thus to trace an axonal process across 1 mm requires tracking that axon across each of ~10^5 individual sections. An error in even a single section raises the possibility that a particular axonal process will be assigned to the wrong parent cell body. Although there has recently been impressive progress improving its accuracy and throughput, electron microscopy remains an inherently challenging approach to connectomics, particularly for studying long-range connections such as those to and from the thalamus, striatum, and other cortical regions.

However, until recently there was no alternative to EM for solving the connectome at single neuron resolution.

To circumvent the considerable challenges associated with EM-based connectomics, my laboratory is developing an entirely novel class of approaches called BOINC (Barcoding Of Individual Neuronal Connections). BOINC relies on high-throughput DNA sequencing, a technology originally developed to sequence genomes from humans and other organisms. The appeal of using sequencing is that its cost is plummeting: it is now possible to sequence an entire human genome (~three billion nucleotides) for a bit more than $1,000, compared with $1 million in 2007 (for James Watson's genome), and more than $2 billion for the human genome project (2001). Indeed, the cost of sequencing is falling at a rate that exceeds even Moore's law, which states that computer power doubles every two years. DNA sequencing has not previously been proposed for connectomics, but we reasoned that if we could convert neural connectivity into a sequencing problem, we could render it tractable using current, low-cost techniques.

We are pursuing several strategies for converting connectomics into a sequencing problem, but all BOINC methods must solve three challenges. First, we must express a unique sequence of DNA—a DNA "bar code"—in each neuron in the brain. Since DNA consists of long strings of four nucleotides (A, T, G, C), a bar code consisting of a random string of thirty nucleotides can uniquely label $4^{30} = 10^{18}$ neurons, far more than the number of neurons in the mouse cortex ($< 10^7$ neurons). Thus the vast majority of neurons will have a unique bar code. Second, we must induce each neuron to share copies of its bar code with its synaptically coupled partners. Finally, we join pre- and postsynaptic bar codes into a single molecule suitable for high-throughput DNA sequencing. The presence of a joined pair of pre- and postsynaptic bar codes indicates that those two neurons are connected. It is thus straightforward to fill in the entries of the (sparse) connection matrix directly from the observed bar code pairs.

The first challenge is to bar code neurons. The most appealing solution is to make a transgenic mouse engineered with a genomic cassette—a specific sequence of DNA inserted into a known location on a chromosome—that is scrambled randomly in each neuron. This cassette would consist of special short sequences of DNA termed "recombinase sites," S, which flank intervening sequences $X_1, X_2, \ldots X_N$ (in which each X is used to denote a short sequence of nucleotides like X = AAGGCCCCATTA). The transgenic mouse would also be engineered to transiently express a special protein, termed a "recombinase," which inverts the DNA between a pair of recombinase sites. Thus in one neuron the original germ-line sequence $S\,X_1\,S\,X_2\,S\,X_3\,S$ might be scrambled to produce $S\,x_3\,S\,X_1\,S\,x_2\,S$ (where lowercase letters are used to denote the inverse sequence, x = ATTACCCCGGAA for the example above), whereas in another neuron scrambling might generate $S\,X_2\,S\,x_3\,S\,X_1\,S$. The theoretical diversity D achievable by this strategy grows rapidly with the number of intervening sequences N as $D = 2^N N!$, which is the number of sequences of N playing cards, assuming cards are not only shuffled but can also be flipped face up or down. Although such recombination, or scrambling, may seem fanciful, it is in fact analogous to the mechanism by which antibody diversity is generated in the vertebrate adaptive immune system. Recombination solves the problem of how to endow individual cells—all derived from a single egg and therefore by default genetically identical—with unique sequences.

The second challenge is to share bar codes between synaptically connected neurons. We have previously outlined an approach to this based on Pseudorabies virus (PRV), a member of the Herpes virus family (Zador et al., *PLOS Biology*, 2012). PRV, like all viruses, is essentially a core of genetic material (DNA in the case of PRV) wrapped in a protein coat. Unlike most viruses, however, PRV propagates from neuron to neuron across the synaptic cleft. PRV evolved this method of propagation (which it shares with rabies virus, to which it is otherwise unrelated) as a way of penetrating the nervous system while evading immune surveillance. Because PRV moves efficiently across synapses, neuroscientists have long used it to trace neural circuits; tracing studies often use an attenuated form of PRV that can only propagate in the retrograde direction. For BOINC, we engineered the genetic material in this PRV to include a bar code, so that a neuron passes its bar code to each of its synaptically connected partners. Thus each neuron becomes a bag of bar codes, a bag containing copies of its own bar code as well as of its synaptically connected partners.

The third challenge is to join bar codes within a neuron. To achieve this we express a specialized protein called an integrase. Like the recombinase described above that inverts DNA flanked by recombination sites, the integrase also acts upon pairs of integrase sites. However, the integrase irreversibly joins the DNA at the sites, forming a single piece of DNA out of two. By positioning the bar code sequences near the integrase sites, we ensure that the single piece of DNA thus formed contains two bar codes in sequence. This single piece of DNA can be amplified by conventional methods and sent for high-throughput sequencing.

BOINC has two key advantages over electron microscopy. First, it is much cheaper. Given current (2013) costs, a mouse cortex with $< 10^7$ neurons and perhaps 10^{10} synapses might require a few weeks and about \$10,000 to sequence, numbers that will only improve as sequencing technology advances. Second, BOINC is particularly well suited to studying long-range projections because the error rate does not increase with the length of the projection. Thus BOINC can be used to study not only local circuitry within a cortical module but also its long-range connections.

Two limitations of BOINC in its simplest form are that (1) it has no natural representation of space, so that the identity of a bar code provides no information about its spatial position in the circuit (for example, auditory versus visual cortex), and (2) it has no natural representation of cell type, so that the identity of a bar code provides no information about whether the associated neuron is, for example, excitatory versus inhibitory. We can address the first concern by keeping track of the brain area from which each bar code is obtained at the time of dissection, prior to extracting the nucleotide bar codes. The spatial resolution here can be as low as 100 μm or even lower, sufficient to assign each bar code to a defined anatomical region. We address the second concern by barcoding not only synaptic connections but also the "transcriptome" associated with a given neuron. The transcriptome is the collection of mRNA (messenger RNAs) transcripts that couple a cell's DNA to the proteins it expresses. These mRNAs define whether a neuron is excitatory or inhibitory as well as provide other information such as the cortical layer from which it was obtained. Thus we envision a connection matrix in which associated with each neuronal bar code are a few additional bits of information specifying the neuron's position in the circuit and its identity.

A cheap and rapid method for deciphering the wiring diagram of a neural circuit or of an entire organism would have a profound impact on neuroscience research. Many neuropsychiatric diseases such as autism and schizophrenia are thought to result from disrupted neuronal connectivity, but identifying the disruptions even in mouse models is a major challenge given current technology. More fundamentally, knowledge of the neuronal wiring diagram would provide a foundation for understanding neuronal function and development in the same way that knowing the complete genomic sequence provides the starting point for much of modern biological research in the postgenomic era. BOINCing may not solve the brain, but it promises to bring us one step closer.

ROSETTA BRAIN

George Church

With Adam Marblestone and Reza Kalhor

The Multileveled Complexity of the Brain

As with many biological systems studied in the past, the more we look at the brain, the more we find complexity. To start, the neurons are packed densely in a 3D matrix with upwards of 100,000 neurons and 900,000,000 synaptic connections per cubic millimeter of brain tissue. Moreover, neurons come in hundreds (or perhaps thousands) of functionally distinct cell types with unique morphologies and molecular (epigenetic) identities.

Synaptic connections can be excitatory or inhibitory and can transmit information using more than one hundred distinct neurotransmitter molecules. These connections change strength, break, and reform over time, and can even alter which neurotransmitters they use in response to experience. Furthermore, gaseous messengers (which permeate indiscriminately across cell membranes) and long-range electrical interactions may allow communication beyond the confines of yesterday's chemical and electrical synapses.

Neurons are not the whole story: other cells like glia, once viewed as mere metabolic support infrastructure, are now thought to play important roles in dynamic information processing. For example, neurons make synapses onto glia, and glia release neurotransmitters that modulate information flow between nearby neurons.

Going deeper, each cell (whether neuron, glial cell, or otherwise) is composed of a network of self-assembling molecular machines, the dynamics of which is used not only to *construct* the electrochemical computing elements (neurons) but also to *dynamically store and manipulate information* within genetic logic circuits and synaptic protein assemblies.

On the other hand, the fully functional brain self-organizes from a less-structured precursor during development and learning. If we understood the rules governing these self-organization processes, we could begin to know which aspects of the brain's complexity are relevant or irrelevant, at different scales and in different types of neural computation. The apparent complexity will likely continue to pile on, however, at least until we understand the principles underlying brains and minds much better, so that we know something about what to look for and what to expect. This presents a "chicken and egg problem" for our present moment in neuroscience; we must look ever-more comprehensively at the brain's complexity in order to have hope of understanding any deeper simplicity it may possess—not to mention fixing its ailments and interfacing to it with appropriately high bandwidths.

Approaches to Comprehensive Brain Mapping and Modeling

Recently initiated large-scale efforts in neuroscience have focused on three projects: connectomics (mapping which neuron is synaptically connected to which others; see chapters by Sporns, Zador, and Hawrylycz herein), brain activity mapping (observing the electrical "traffic" along these "synaptic highways," see chapters by Shenoy and Koch), and large-scale brain simulation (integrating data from all areas of neuroscience to construct biophysically realistic models that can be compared with experiment, see chapter by Hill). While each of these endeavors is extraordinarily valuable, none on its own is matched in scope to the brain's multileveled complexity. Furthermore, it is not always obvious how to put these projects together, or easy to do so in practice, since each has missing pieces that the others don't make up for.

For example, having an activity map without a connectome could tell us much about emergent behaviors across large neural networks but might not be sufficient to reconstruct the underlying circuitry. Having a connectome would be informative as to circuit architecture, but would not necessarily specify the excitatory or inhibitory nature of synapses and would represent only a static picture, leaving mysterious the rules of development and plasticity that construct the circuits in the first place (acquiring many connectomes at different time points could help, but this would be difficult to do for a *single* animal). Simulations are powerful as a means to integrate diverse types of knowledge into predictive models that can be compared with experiment, but without data even more fine-grained, comprehensive, and multifaceted than activity maps *and* connectomes, these simulations may be too underconstrained to reflect the important aspects of the brain's functional architecture.

What Might the "Right" Dataset Look Like?

Before we despair in the complexity and give up on the brain entirely, it behooves us to ask a childlike question: regardless of apparent feasibility, what dataset would we ideally like to have to help understand how the brain's structural and functional biological levels interlink to form an integrated system?

At a minimum, we would like to observe, simultaneously—within a *single* brain—information that reports on *all* of the levels of complexity that we outlined above. As a start, we could imagine a dream experiment that would report on:

Cell types
Connections
Connections strengths and types
Developmental lineages
Histories of electrical activity patterns over time
Histories of molecular changes over time

To imagine what this dataset would represent, an analogy to a very different field is useful. The Rosetta Stone is a 1,700-pound tablet bearing three inscriptions, carved one above the other into the stone: a priestly decree in honor of King Ptolemy V in ancient Egyptian hieroglyphics, in ancient Greek script, and in demotic script.

Because it presented the exact same statement in three different languages, two of which were known and the third unknown, it provided a key resource for cracking the then-unknown code of the hieroglyphics. Similarly, a *Rosetta Brain* would convey information about multiple interrelated phenomenological levels of the brain's biology and allow these levels to be directly compared to one another with single-cell precision.

Thus each neuron in a Rosetta Brain would report a record not only of its own electrical activity pattern and of its connectivity but also of its developmental lineage, cell type, and history of synaptic and ion-channel protein concentrations.

Abstracting the Problem

In the rest of the chapter, we suggest a way in which all this might be possible. Our approaches start with the fact that all of the observations we wish to make across the different levels of a Rosetta Brain come down, in essence, to operations of *labeling* and *counting*. For example, a connectome is at its core a gigantic matrix, specifying whether or not cell X is synaptically connected to cell Y, where X and Y may be any of 100,000,000 neurons (in the mouse brain). As Tony Zador and colleagues proposed (see the chapter on sequencing the connectome), if each cell had a unique name (a unique string of letters), which we can also think of as an ID card or bar code, then for each name string, we would merely need to ask whether

(name string #1, name string #2)

is within the list of known connections, to find out if the corresponding synaptic connection is present.

Developmental lineages of neurons are equally simple, in concept; just give each cell a unique bar code, but such that the children of cell X have bar codes of the form

child(bar code of cell X)

and those cells' children have bar codes like

child(child(bar code of cell X))

and so on.

The many neural cell types, although traditionally defined by the complex morphologies of axons and dendrites first glimpsed under the microscopes of Camillo Golgi and Santiago Ramón y Cajal, can also be defined by a process of discrete counting. Indeed, all of the cells in the body share (very closely) the same genome, the differences between them being due to the different levels of expression of the different genes. From the Central Dogma of Molecular Biology we know that gene expression, which accounts for the phenotypic differences between the body's genetically identical cells proceeds as

DNA →(transcription) Messenger RNA →(translation) Protein

and thus, *by counting the numbers of each messenger* RNA in a cell, we can determine its cell type.

The same goes for tracking the history of molecular expression in the cell over time, for example, to observe changes in gene expression that accompany learning and memory; except here you need not only to count molecules but also to label these molecules with time stamps— digital strings that encode the current time. This would be similar to how a grocery store tracks its sales: every time a bar-coded item is scanned at the checkout counter, the time is also recorded, and the time-stamped bar codes are entered into a database.

It is less obvious how connection strengths and types can map into this language, but in principle, these could be inferred by counting the abundances of different proteins on either side of the synapse, since the distribution of neurotransmitter receptors and other synaptic proteins ultimately determines the nature of the synapse. A further variable that influences connection strength is number of distinct synapses (axon terminal to dendritic spine contacts) made by cell X onto cell Y. Thus a means to count individual synapses would provide a crude index of connection strength.

There is yet another way in which connection strengths could be determined, however, if we had access to another level of the Rosetta Brain: electrical activity histories. If we had the full electrical activity histories of cell X and cell Y at a sufficiently high temporal resolution, it would be possible to "see," in these time traces, the moments at which an electrical impulse (also known as an "action potential" or "spike") from cell X is transmitted across a synapse to generate an impulse in cell Y shortly thereafter (in cooperation with inputs from many other cells that synapse onto cell Y). By tracking the statistics and relative timings of impulses from cells across the network, one could determine an effective "functional connectivity" between every pair of cells. Furthermore, by combining this functional connectivity information with information about the underlying anatomical connectivity matrix, it might be possible to compute the actual synaptic strengths between synaptically adjacent cells. In effect, by combining sufficiently rich, redundant, and interconnected datasets, it may be possible to "fill holes" in any one such dataset. Although this leads to highly nontrivial statistical problems, progress is already being made on reconstructing anatomy from activity in the smallest neural circuits.

Sequence Space: An Exponential Resource Matched to the Brain's Vastness

We've conceptually reduced the problem of constructing a Rosetta Brain to massive repetition of a simple operation: reading (and counting) of "bar codes" or "labels." We've seen that if each cell, each synapse, or each molecule could have a unique bar code—and ideally a bar code that also encodes a time stamp—then by counting these bar codes and correlating them with independent measurements of the cell's electrical activity history, we could potentially combine these data to infer an enormous amount about the brain's structure and dynamics. But how do you generate and read "bar codes" at the subcellular level?

That's where DNA comes in. Although we are taught in school to think of DNA as the medium in which the cell stores its genome (its genetic blueprint and instruction set), the capabilities of DNA as an information storage molecule go far beyond that. Indeed, DNA can take on any sequence of the four chemical letters: A, T, C, and G (for instance, ATATAGATAGATCACCCAGAAGATAGGAT is a perfectly valid string of DNA). This simple observation—that DNA can store any string, not just those used as biological blueprints in the genomes of existing organisms—has startling implications for many areas of science and technology because it provides us with a strategy to extend information technology to the molecular level and to integrate it with biological systems.

Meanwhile, sequencing technologies developed in academia and industry have put DNA sequencing on a cost-performance trajectory that outpaces Moore's law, which governs the improvements in silicon microprocessor technology that have brought us from arm-sized cellular phones to Google Glass in only two decades. Many of the same concepts have been adapted to DNA synthesis, which is now on a similar trajectory. This has resulted in the ability to read and write information into DNA with unprecedented ease, as demonstrated recently by the 2012 DNA encoding and subsequent reading of the text of a complete book (*Regenesis*, Basic Books).

Suppose we have a string of DNA twenty-five letters (deoxyribonucleotides) in length. How does the number of distinct DNA sequences of this length compare with the number of synapses in the brain?

$$\text{\# of DNA sequences of length 25 nucleotides} = 4^{25}$$
$$\text{\# of synapses in human brain} = 10^{14} \sim 4^{23}$$

Thus the number of possible DNA sequences of length 25 exceeds the number of synapses in the human brain by a factor of nearly 100. Furthermore, it is easy to make a test tube with all 4^{25} possible DNA sequences of length 25. First, mix together all four letters (A, T, C, and G) so that they react to form all pairs (AA, AT, AC, AG, TA, TT, TC, TG, CA, CT, CC, CG, GA, GT, GC and GG). Then take these pairs and add in all four nucleotides to make all triplets. Repeat this 25 times and you have all possible DNA sequences of length 25 in your test tube.

Now take a test tube with many copies of all these random DNA sequences, which we can also call "DNA bar codes," of length 40 (length 25 is cutting it a bit too close for comfort), and suppose that one could randomly insert one such sequence in each of the roughly 100 million (10^8) neurons in the mouse brain. Then what is the likelihood that two mouse neurons end up with the *same* bar code? This is mathematically identical to the famous Birthday Problem: in a group of k people, what is the likelihood that any two people share the same birthday (among n = 365 possibilities). Here $k = 10^8$ and $n = 4^{40}$, and it turns out that the chance that two neurons end up with the same DNA bar code in this scenario is less than 1 in 100,000,000,000.

Thus by supplying a random DNA bar code to each cell in the mouse brain, we can give each cell a unique label. Similarly, the trick to carrying out all the labeling and counting operations needed for our Rosetta Brain is to encode as much of the relevant information as possible in a

DNA form. But how, in practice, can we read these DNA sequences in the context of an intact brain?

In situ Sequencing: A Key Tool for Rosetta Brain

Normally, when we sequence DNA in the lab, the DNA starts out as molecules freely diffusing in a clear liquid in a test tube. We put the test tube into a machine, and out comes a long text file with line after line of DNA letters and associated metadata.

The project we describe would require the comparable processes to happen inside brain slices. In a sequencing machine, the individual DNA molecules are randomly deposited on a flat glass surface (like a microscope slide), where they are then trapped in place. Next, DNA polymerase, which makes copies of DNA, is added to the reaction. DNA polymerase builds up a copy of the chain from freely floating individual letters (A, T, G, and C). It thus spits out many identical copies of each DNA molecule on the surface, which become trapped right next to it, forming a cluster or colony of identical DNA molecules at a particular spot, which can be seen under a microscope. Then another copy is made, except this time, chemically modified versions of A, T, G, and C are used, each of which is attached to a different color of fluorescent dye: A-red, T-green, G-blue, C-yellow. Thus when DNA polymerase makes a copy of the DNA strands in a particular colony, moving letter by letter along the chain, the colony will show up in red under the microscope when A-red is added to the chain, and similarly for the other colors. By recording the changing colors of the spots, the machine—which is basically a microscope plus some plumbing to pump in the A, T, G, C, and polymerase at the right times—can read the sequences of DNA molecules all across the glass surface at the same time. This mode of doing DNA sequencing with a microscope is part of what has allowed sequencing technology to become so cheap—because the microscope can see many colored spots at once at different positions on the surface.

In a postmortem brain, which we have removed from the animal and sliced into thin slices so that we can look through each slice with a microscope (a thin enough slice will be largely transparent), we could do something comparable, beginning by applying chemicals to "fix" the

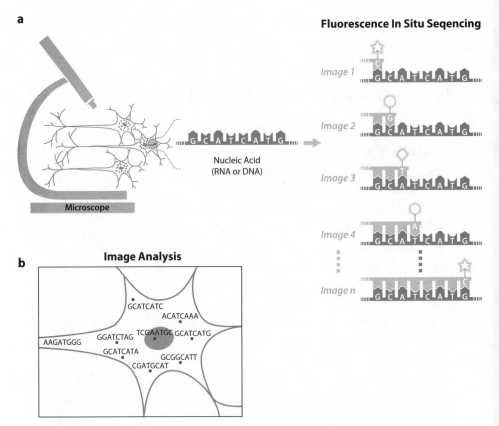

Figure 1. Fluorescent In Situ Sequencing (FISSEQ): a. Sequencing by synthesis: a microscope records the changing colors of DNA spots as fluorescent DNA letters are incorporated into growing chains by a polymerase. Each spot is made of many copies of a single "parent" molecule. b. This results in a set of identified points, each labeled by its corresponding sequence.

tissue so that it does not degrade over time and is mechanically rigid. Then, instead of doing our sequencing-by-microscopy on a clean glass surface with DNA strands *deposited* on it, we can do sequencing-by-microscopy on a *slice of brain with DNA or RNA strands already inside it*! We call this new technology Fluorescent In Situ Sequencing (FISSEQ) because it uses a fluorescent microscope to do DNA sequencing using colored nucleotides "in situ," or in other words, inside an intact slice of brain tissue.

Applying In situ Sequencing to Determine Cell Type, Connectivity, and Lineage

As in situ sequencing continues to improve, we will possess a powerful means to create an "annotated connectome," or in other words, a connectivity map where the cell type of each neuron is known. To do this, we need three things:

1. Delivery of a unique DNA bar code to each neuron. This could be done using random DNA sequences as described above, shuttled to each neuron and inserted into its genome by using a harmless (genetically modified) virus.
2. A fluorescent label, which specifically sticks to synapses, to allow us to see in the microscope where the synapses are located.
3. An in situ sequencing microscopy setup with high spatial resolution (the synapses are packed so densely that only a few wavelengths of visible light can fit in between, perhaps requiring the use of "super-resolved" light microscopies).

To determine connectivity, we can then look in the microscope for the locations of synapses and apply in situ sequencing to read the sequences of the nearby bar codes on either side of the synapse. This will tell us which cell bar codes are paired with which other cell bar codes across synapses. (See chapter by Zador for a more sophisticated version of this scheme, where viruses are used to shuttle DNA bar codes between synaptic partners—this could enable DNA sequencing to be used to read out a connectome using commercially available sequencing technologies [as opposed to the emerging in situ technology]! On the other hand, by directly sequencing the bar codes in a microscope using the in situ methodology, we would avoid the need to shuttle bar codes from cell to cell across synapses, as is required in Zador's approach.)

To determine the cell type annotations, we can directly apply the in situ sequencing approach to the messenger RNAs inside each cell, which will provide us with a "profile" or "pattern" of its gene expression, which is a good indicator of the cell type. To determine cell lineage in addition to cell type, we need DNA bar codes that change slightly every time a cell division occurs. Then, by tracing back these small changes, we can determine the "family tree" of each cell. This is similar to the way researchers already use DNA sequencing (applied to genomes, not bar codes) to determine the family trees of actual human families, except applied to different cells within a *single* brain.

Approaches using electron microscopy (which allows roughly 100x higher resolution than standard light microscopy can provide) also have the ability to combine functional studies with detailed probing of the underlying circuit connectivity. In these "EM Connectomics" approaches, connections and cell types must be inferred from the high-resolution microscopy images, because electron microscopy is less easily combined with multicolor molecular reporters and DNA sequencing. This poses challenges since axons must be tracked over large distances inside huge image stacks and synaptic connections, and cellular morphologies must be computationally inferred from high-resolution image data. Doing so requires slicing the brain into nanometer-thin slices, which means that the density of data in all three dimensions is higher than would be required in an optical approach. EM Connectomics is extremely powerful, and remarkable progress is being made on both hardware and image analysis, as demonstrated by recent complete-circuit reconstructions of retinal circuits in the fly and mouse. Yet because of the exponentially rich information-encoding capacity of DNA, and its facile readout through sequencing, we would suggest that a Rosetta In Situ Sequencing approach may have complementary features, particularly in that it naturally integrates multiple forms of data.

In situ Immune Microscopy of Synaptic Proteins to Determine Synapse Strengths

So far, we haven't specified a great way to obtain the synaptic strengths and types. Fortunately, with the same microscope used for in situ sequencing, we can apply well-known techniques for visualizing synaptic proteins, which rely on special forms of molecular recognition called antibodies to bind a specific color to a specific synaptic protein. Because the distribution of synaptic proteins is an indicator of the strength and type of the synapse, we can use this method in concert with in situ sequencing to further annotate the connectome with synaptic parameters.

We could even combine this antibody staining approach with in situ sequencing by linking DNA strands to specific antibodies. This would effectively give us the ability to simultaneously read out 4^N colors instead of just 4 colors (just as we can with RNAs already, see above).

Encoding Neural Electrical Activities into DNA?

What about time-dependent phenomena, most importantly the rapidly varying electrical activity of each neuron—a.k.a. the activity map. Is it possible to read this out from in situ DNA sequencing as well? Although it might sound unlikely to map between dynamic cellular electricity and static strands of DNA, we can foresee at least one way of doing so.

Imagine again a DNA polymerase copying a long DNA strand. To do so, it works its way from one end of the strand to the other, effectively reading the identity of each nucleotide along the chain and then grabbing, from solution, the complementary nucleotide to form the next link in the strand representing the copy. Now imagine that we could "mess up" this copying process just for an instant, so that mistakes would be made and the wrong letters incorporated into the chain. If we knew when the polymerase started going at one end, we could track approximately *when* this perturbation occurred by looking at the position along the chain where the cluster of mistakes occurred. If the perturbation happened later, the polymerase will be farther along the chain, and thus the cluster of mistakes will occur farther out along the chain from the starting point.

Now imagine if we could cause the polymerase to make more or less copying mistakes in response to the instantaneous level of neural activity. Then the pattern of mistakes along the DNA chain would be like a "ticker tape" record of the pattern of neural activity in time. One potential way to make this happen relies on the fact that when neural electrical activity occurs, calcium ions rush into the cell. These calcium ions could make their way to the polymerase and disrupt its fidelity of copying, causing more mistakes to occur in the presence of high calcium.

Although there are many challenges that must be overcome to get such molecular ticker tapes running in the lab, this idea has already spawned intellectual descendants that may be easier to implement; for instance, to record events on slower timescales into a DNA storage medium for later

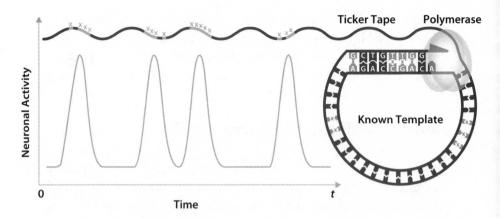

Figure 2. Molecular recording devices: the dynamic flow of ions across the cell membrane during neural activity could be recorded into DNA by modulating the copying mistakes made by a DNA polymerase. This would allow readout of the activity history by DNA sequencing after the fact, so that real-time access to each neuron by external devices is not required during recording.

readout by sequencing. It will then be possible to record time-varying histories of molecular events in the cell into a static medium, DNA, and then to read out those histories using in situ sequencing.

Importantly, even before these technologies are in place, we can also read out the electrical activity patterns for at least a small number of cells with already-extant methods, such as using wired electrodes to directly sense the electrical voltage associated with neural impulses. Then, the same brain used in these experiments can be subject to the Rosetta in situ sequencing procedure to read out other salient aspects of its structure and dynamics.

Architecture for a Rosetta Brain Experiment

Putting this all together, we can imagine an experiment as follows.

Part I. Living animal

- Deliver appropriate DNA bar codes or other molecular markers
- Do as many experiments on a behaving organism as possible

- Stimulate as much as possible
- Record electrical activity from as many neurons as possible in real time, via traditional methods

Part II: In situ

- Chemically fix the brain and slice through the head to create transparent sections
- Perform in situ sequencing and in situ microscopy of synaptic proteins
- In this final step, get as much information as possible about the nervous system via in situ microscopy and sequencing: cell types, developmental lineage bar codes, connectome bar codes, antibody staining of synaptic proteins, and ideally molecular ticker tape activity time-series data.

The Future

Obtaining the massive interrelated datasets resulting from a Rosetta Brain would be just the first step. It should be made easy enough to create a Rosetta Brain that many small labs could make their own, under different experimental conditions designed to test a wide range of influences and hypotheses. Rosetta Brains should be compared between individual animals to understand how brains vary and what they have in common. Systematic approaches would also be possible—Rosetta Brains would be ideal as datasets to compare against putative large-scale brain simulations. They could be used to ask many questions at once, in a real brain, as is already possible in a computational brain model, and to probe how each variable is related to many others.

Further Reading

Kording, Konrad. 2011. "Of Toasters and Molecular Ticker Tapes." *PloS Computational Biology* 7 (12): e1002291. http://www.ploscompbiol.org/article/info%3Adoi%2F10 .1371%2Fjournal.pcbi.1002291.

Lee, Je-Hyuk, Evan Daugharthy, Jonathan Scheiman, Reza Kalhor, Richard Terry, Joyce L. Yang, Chao Li, Ryoji Amamoto, Derek Peters, Thomas C. Ferrente, Adam Marblestone, Amy Bernard, Brian M. Turczyk, Nicholas Conway, Samuel Inverso, Daniel Levner, Prashant Mali, Xavier Rios, Sauveur S. F. Jeanty, Allan R. Jones, John Aach and George M. Church. 2014. "Highly Multiplexed Three-Dimensional Subcellular Transcriptome Sequencing In situ." *Science*: dx.doi.org/10.1126/science.1250212.

Marblestone, Adam H., Bradley M. Zamft, Yael G. Maguire, Mikhail G. Shapiro, Thaddeus R. Cybulski, Joshua I. Glaser, Dario Amodei, et al. 2013. "Physical Principles for Scalable Neural Recording." *Frontiers in Computational Neuroscience* 7. http://www.arxiv.org/abs/1306.5709.

Marblestone, Adam H., et al. 2014. "Rosetta Brains: A Strategy for Molecularly-Annotated Connectomics." *arXiv preprint arXiv*:1404.5103. http://arxiv.org/abs/1404.5103.

Zador, Anthony M., Joshua Dubnau, Hassana K. Oyibo, Huiqing Zhan, Gang Cao, and Ian D. Peikon. 2012. "Sequencing the Connectome." *PLoS Biology* 10 (10): e1001411. http://www.plosbiology.org/article/info%3Adoi%2F10.1371%2Fjournal.pbio.1001411.

COMPUTATION

Even with a map of every neuron and every connection in the brain, we will only be part way there. A road map would tell us a lot about the United States—it would be clear that New York City and Chicago are major hubs, and that there are roads of different magnitude, and so forth. But there is much that could not be inferred from maps alone (even if traffic were superimposed). Where some organs (like the liver or the nose) make sense almost immediately once we understand their constituent parts, the brain's operating principles continue to elude us.

In part, that's because the brain is an organ of computation: whereas liver cells remove toxins, and the nose filters out pollutants, nerve cells exist to compute; the real trick is to figure out *what* they are computing. And, to use an analogy from computers, it is as if we are trying to reverse engineer the operating system, the software, and the protocols and conventions (like USB, the ASCII code, and TCP/IP) for communication—all at once. The challenge is enormous.

In this section, we'll hear about how some leading researchers are approaching the challenge of understanding the brain's computations. We begin with the work of **May-Britt Moser** and **Edvard Moser**, who are dissecting the neural circuitry underlying spatial navigation. Their work provides a detailed link between the anatomy and physiology of a particular brain area and the computation it appears to be performing; it is a paradigm of what we hope the field of neuroscience will be able to achieve throughout the brain. **Krishna Shenoy** describes a philosophy of data analysis that tries to see the forest through the trees: it is one thing to analyze a neuron or two at a time, but how we do begin to understand the interaction of thousands or even millions of neurons? **Olaf Sporns** emphasizes the role of large-scale networks in the brain and argues that the mathematics of collective and emergent behavior can provide important insight into neuroscience and a framework for interpreting the big data that will emerge. **Jeremy Freeman** describes the onslaught of data that is coming in neuroscience. He explains how new technologies may help us handle it, but also why truly understanding the brain will require more that just knowing what to do with the data.

UNDERSTANDING THE CORTEX THROUGH GRID CELLS

May-Britt Moser and Edvard I. Moser

One of the ultimate goals of neuroscience is to understand the mammalian cerebral cortex, the outermost sheet of neural tissue that covers the cerebral hemispheres. All mammalian brains have a cortex, but during evolution, the size of the cortex has expanded enormously, and in the largest brains the growth has resulted in extensive folding, with much of the cortical surface getting buried in deep grooves, or sulci and fissures. The cortex is the site where most cognition and intellectual activity takes place. Thinking, planning, reflection, and imagination depend on it. Memories are stored there, and the cortex takes care of language interpretation as well as language production. Moreover, although the cortex can be found across the whole range of mammalian species, the expansion of this brain structure is thought to underlie the amplification of the intellectual repertoire in humans.

Can We Understand the Cortex?

What is the neural basis of the intellectual functions of the cortex? At its outer limits—such as in deciphering the neural basis of higher brain functions—the cortex may at first seem quite unreachable. But it is important to remember that the cortex also performs more tangible operations, such as interpreting inputs from the sensory environment. The cortical interpretation of sensory signals has for a long time served as the neuroscientist's window into the cortex. By studying, for example, how signals from photoreceptors, olfactory receptors, or touch receptors are represented at early stages of brain processing, in the primary sensory cortices, neuroscientists have made quite significant progress in describing and understanding some of the operational language of the cortex.

A major breakthrough in the analysis of sensory cortices was the discovery of cells that responded selectively to local features of the visual field. In a series of experiments that started at the end of the 1950s, David Hubel and Torsten Wiesel showed that neurons in the primary visual cortex fired specifically when line segments of a specific orientation were presented to the visual field. The linear receptive fields of these cells differed from the circular center-surround fields of cells at earlier stages of processing, in the retina and in the thalamus. Hubel and Wiesel showed that cells with different orientation preferences, or different preferences for inputs from the left and right eyes, were organized in columns of cells with similar functional properties. These findings pointed to a functional architecture for visual computation in the primary visual cortex and provided unprecedented insight into how visual input was fragmented and reassembled at different stages of the visual system and how function was divided across different elements of the visual circuit. Their work started a new era of neuroscience in which the visual cortex served as a guide to cortical computation, with an impact far beyond the direct implications for the mechanisms of vision. Similar progress has subsequently been made in other sensory systems, and we are beginning to understand how senses as diverse as olfaction, taste, and touch are encoded at the level of cortical circuits.

However, while remarkable insights have been made at the early stages of cortical processing, where the first transformations of sensory input take place, little is known about how the brain works at subsequent levels, in the higher-order integrative parts of the cortex—the "high-end cortices." Yet this is probably the territory of the most challenging cognitive operations, such as the production of thought, decisions, or complex memories. One of the reasons for the inaccessibility of the higher parts of the cortex is that as the distance from the sensory receptors increases, the firing of the neurons becomes increasingly decoupled from the specific features of the sensory environment. As such, it becomes difficult to find correlates in the external world that possess any predictable relationship to the firing pattern of the recorded cells. At the high end of the cortical hierarchy, firing may be triggered via a multitude of converging sensory channels as well as intrinsic processes not corresponding to any particular input. When we do not know the active inputs to a cortical area at any given time, and those inputs originate from areas whose workings we also do not understand, it is difficult to relate the activity of a cell from one of the high-end cortices to any particular behavioral function.

The Mammalian Space Circuit: A Window to the High-End Cortices

One exception to the apparent decoupling from the external world is the space-encoding cell population of the hippocampus and the entorhinal cortex, located at the very top of the cortical hierarchy, many synapses away from any of the primary sensory cortices (see color plate 4). In this system, cells have remarkably predictable firing correlates. Many cells in this part of the brain fire only when the animal is in a specific set of locations in its local environment. The preferred locations differ from cell to cell, such that as a population, the cells fire in unique combinations at every location in the environment. Because of these unique activity combinations, the cells effectively serve as a map of the animal's position.

The study of the neural basis of space began in 1971, when John O'Keefe and John Dostrovsky, at University College London, used microelectrodes to record natural activity of neurons in the hippocampus of freely moving rats (figure 1). They were able to pick up impulses, or action potentials, from individual cells in CA1, one of the major subfields of the hippocampus. Many of their cells responded specifically to the animal's location in the environment. These cells were named "place cells." When the rat was in the cell's "place field," the cell fired at a high rate. As soon as the rat left this area, the activity decreased and remained low until the next time the animal came to the place field. O'Keefe and his colleagues soon discovered that most hippocampal cells had place fields and that the exact firing locations differed from one cell to the next. Collectively the population of place cells was found to generate a map of the environment, with a unique constellation of active cells at every single position. The strict relationship between neural activity and a property of the environment—the animal's location—was unique among all the recordings that had been made in higher-end cortices by that time.

Figure 1. Most of our knowledge of the mammalian space circuit has been obtained from rats and mice. Rodents have a well-developed entorhinal and hippocampal cortex and demonstrate excellent spatial memory and navigation—skills thought to depend on these cortical systems.

During the decades following the discovery of place cells, accumulating evidence suggested that place cells have functions that extend beyond a specific role in the mapping of the physical space. This idea was reinforced by the observation that different maps could be activated by small variations in the appearance of the environment, suggesting that if the brain has a general map for distances and directions that does not care about what the environment looks like, then such a map should be located elsewhere. Motivated by these considerations, we started, at the turn of the millennium, to search for spatial representations outside the hippocampus. In our first study, with Vegard Brun and a few other graduate students, we recorded from CA1, the hippocampal subregion where place fields had been identified thirty years earlier. In some of the animals we removed the intrinsic connections of the hippocampus, leaving intact only the direct input from the entorhinal cortex, upstream of the hippocampus. Somewhat to our surprise this interference with local hippocampal circuits did not abolish the tendency for CA1 cells to fire in specific locations. Place cells remained place cells. This implied that unless the place signal was generated entirely by local CA1 processes, the cells must have received critical spatial input from the entorhinal cortex. The findings drew our interests to this unexplored cortical region, one synapse upstream. This brain area turned out to be a gold mine.

Grid Cells and Grid Maps

In 2004 and 2005, together with our students Marianne Fyhn, Torkel Hafting, and Sturla Molden, and our colleague Menno Witter, we inserted recording electrodes directly in the medial entorhinal cortex, in a part of the area that was strongly connected to the locations in the hippocampus where place cells were normally studied. The findings were quite striking. Individual cells had discrete firing fields, like place cells in the hippocampus, but each cell had multiple fields, and the fields were arranged in a remarkably regular pattern (figure 2). Collectively, the firing fields of an individual cell defined a periodic triangular array covering the entirety of the animal's environment, like the cross points of graphic paper rolled out over the surface of the test arena, but with equilateral triangles as the smallest repeating unit, as on a Chinese checkers board. Because of their grid-like periodic firing pattern of these cells, we named them grid cells. The grid structure was similar for all grid cells, but the spacing of the fields, the orientation of the axes, and the x-y location of the grid fields varied. Grid cells were initially observed in rats, then we found them in mice, and more recently they have been described also in bats, monkeys, and humans, suggesting that they are present widely across the mammalian branch of the phylogenetic tree.

A striking property of the grid cells was the persistence of the firing pattern in the presence of changes in the animal's speed and direction. Moreover, when two grid cells were recorded at the same time, the relationship between their grid fields tended to replicate from one environment to the next. If the grid fields of two simultaneously recorded cells overlapped in one task, they would generally overlap in the next too. The rigidity of this relationship was quite different from the behavior of place cells in the hippocampus, which, based on the work of Bob Muller and John Kubie at SUNY Downstate Medical Center, were known to

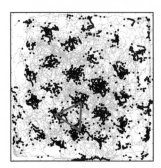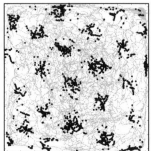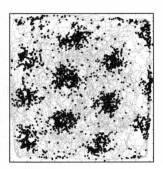

Figure 2. Grid cells in entorhinal cortex of the rat brain. Three grid cells are shown. Left: cell with short spatial wavelength; right, cell with long spatial wavelength. Each panel shows the trajectory of a foraging rat in a 2.2 m wide square enclosure (gray) with the spike locations of one cell superimposed on the track (black). Each black dot corresponds to one spike. Modified from Stensola et al. (2012).

have the capacity to switch between completely uncorrelated firing patterns. The coherent firing patterns of simultaneously recorded grid cells suggested that the same grid map was used over and over again, pointing to the grid cells—and not the place cells—as a likely implementation of a universal brain metric for space.

However, are all grid cells part of the same map, or are there several maps? In the earliest studies, we recorded from few cells at the time, and those cells were generally from the same location in the entorhinal cortex. It was not possible to infer the structure of the grid map from those limited recordings. In more recent work, with Hanne and Tor Stensola in our lab, we have been able to increase the number of neurons by an order of magnitude. By 2012, we were able to record from more than 180 grid cells across widespread regions of the entorhinal cortex in the same animal. These recordings revealed that grid cells are organized into a small number of maps with discrete properties (see color plate 5). Different grid maps varied on a number of parameters, including the spacing of the grid fields and the orientation of the grid axes. At the dorsal end of the entorhinal cortex, near the top, most grid cells had tightly packed grid fields and all seemed to belong to the same module. As we moved away from the dorsal border, cells from other modules, with a larger grid spacing, joined the ensemble, and at the deepest positions, cells with large grid scales often predominated. A total of four grid modules were identified in each animal, but the total number may be larger because only a part of the entorhinal cortex was sampled.

What is most striking about the modular organization of the grid cells is that individual modules can respond quite independently to changes in the spatial layout of the environment. Testing animals with cells from four modules in a square box that was compressed to a rectangle (see color plate 6) showed this. Cells from Module 1—the most dorsal module—did not respond to the compression and kept their original firing locations in the common area of the two boxes. In contrast, in cells from larger modules (Modules 2–4), the grid fields were squeezed in one direction in proportion to the shrinkage of the recording box. These observations suggest that, at least in principle, different modules can respond independently of one another when the geometry of the environment is changed. Apparently the grid network consists of four or more discrete maps that may or may not respond in a coherent manner.

What could be the advantage of organizing the brain's map of space in this way? Why would four or more maps be better than just a single coherent map? The answer may lie in the way the grid map is used "downstream" in the hippocampus. While the majority of the cells in the hippocampus are place cells, the hippocampus is also critical for certain types of memory, often referred to as declarative memories. These are memories that we are conscious of—or can "declare"—including memories of facts and events. Space is a fundamental element of these memories. Because we store thousands of declarative memories every day, the hippocampus needs to find a way to keep them all apart. This is where grid modules may be useful. If two modules respond independently to a change in the environment, their coactivity will change. The change in coactivity will activate a new subset of cells in the hippocampus. Each relative displacement among the grid modules may lead to a different activity combination, which in turn may activate a different set of hippocampal neurons. Thus with only a handful of grid modules, it is possible for the entorhinal cortex to link itself to a large number of hippocampal activity patterns, and putative memories, much like the way a combination lock can store 100,000 codes with only 5 counters that each run from 0 to 9. By combining input from a small number of independent grid modules, hippocampal cell populations may acquire the ability to generate huge numbers of discrete representations individualized to specific places and experiences.

How Are Grid Cells Generated?

One fascinating property of grid cells is that such a regular firing pattern appears so high up in the cortex, far away from the sensory inputs that define the distinct receptive fields of many neurons in the primary sensory cortices. In sensory systems, sensory representations often appear to get more disorderly as the number of synapses from the sensory receptors increases. In contrast, the grid pattern is highly regular, unlike the structure of activity observed so far in areas upstream of these cells. The perfectly hexagonal firing pattern of the grid cells does not correspond to any property of the animal's sensory environment and thus more likely reflects mechanisms that are intrinsic to the entorhinal cortex. What could those mechanisms be—how does a network generate hexagonal firing fields?

While the mechanism of grid formation remains to be established, observations suggest that hexagonal firing patterns emerge as an equilibrium state in competitive networks where all cells inhibit all other cells in their vicinity. Theoretical studies and computational modeling show that in a network where all cells are connected to all other cells within a certain range, via inhibitory connections, hexagonally patterned firing will appear spontaneously as a resting state (see color plate 7). In collaboration with Yasser Roudi and Menno Witter and their colleagues, we have shown that entorhinal cells—in the cell layer that contains the most prototypical grid cells—are connected exclusively via inhibitory interneurons, and that such connections can lead to the formation of hexagonally spaced firing in a model network. Grid cells are perhaps just one of many examples in nature where hexagonal arrangements emerge through self-organizing processes as a result of evenly distributed competitive forces.

Grid Cells Are Not Alone

Soon after the discovery of grid cells in 2005, it became clear that these cells are not the only spatial cell types in the entorhinal network. Grid cells were the most predominant cell type in the superficial parts of the entorhinal cortex, particularly in the cell layer that contains the strong inhibitory connections, but with Francesca Sargolini and other students in our lab, we found in 2006 that a large proportion of the entorhinal cell population is direction modulated. These cells, which are similar to a cell type Jim Ranck and Jeff Taube discovered in other brain regions twenty years earlier, fire selectively when the rat points its head in a certain direction. Some of these cells are grid cells at the same time, firing in grid fields only when the animal moves in the cell's preferred direction. In 2008, with Trygve Solstad, we subsequently found that grid cells and head direction cells intermingle with yet another novel cell type—the border cell. These cells fired specifically when the animal was close to one or several borders of the local environment, such as a wall or an edge. When the box was stretched, the firing field followed the wall, and when a new wall was inserted, a new firing field emerged along the insert. Both head direction cells and border cells retained their properties when the animal was moved to a different environment. Two head direction cells that fired in the same direction in one environment tended to fire in the same direction also in other environments, and two border cells with similar wall preferences in one box would have the same preferences also in another box. The rigidity of the head direction and border cells, as well as the grid cells, suggests that the entorhinal maps are used universally across many environments, much unlike the hippocampal place-cell map, which appears to set up new activity combinations for every single environment or experience.

The presence of multiple spatial cell types in the same neural system, such as place cells, grid cells, head direction cells, and border cells, raises some obvious questions. One is how they are related—are place cells formed from grid cells, border cells, or other cells, and are the entorhinal cells, in turn, dependent on place cells? Recent work by Sheng-Jia Zhang and Jing Ye in our lab has shown that the hippocampus receives projections from a variety of entorhinal functional cell types. The most abundant input comes from grid cells, pointing to these as a major source for place information, but also border cells and even cells with no clear spatial correlate project significantly to the hippocampus. How place cells are generated from these inputs remains an open question, but the observations raise the possibility that place cells receive signals from a variety of sources, perhaps in a redundant manner allowing them to respond at specific locations in response to changing sources of inputs. It is also possible that the functional input to a given place cell varies over time, perhaps with grid cells providing motion-related input at one moment and border cells providing geometric inputs at a different moment. Clear answers to the mechanisms for transformation of signals from one cell type to another will hopefully be obtained during the next few years, considering that experimental tools for addressing such questions are now becoming available.

What We Can Learn from Grid Cells

With the discovery of place cells and grid cells, as well as other spatial cell types, it has become possible to study neural computation at the high end of the cortical hierarchy, quite independently of sensory inputs and motor outputs. A huge benefit of these cell types is the clear correspondence between the firing pattern and a property of the external world—in this case the animal's location in the environment. The presence of an experimentally controllable firing correlate, combined with the access to multiple discrete cell types, makes it possible to determine not only how each of the firing patterns is generated but also how the firing patterns get transformed from one cell type to the next within the network. Grid cells may not only help us understand how representations are generated in high-end cortices, but such knowledge may also feed back to the sensory cortices, where intrinsic and top-down processes may play a greater role than what was previously appreciated.

The space circuit of the mammalian hippocampus and entorhinal cortex is one of the first nonsensory "cognitive" functions of the cortex that may be understood in mechanistic detail within a not too distant future. Understanding how space is created in this circuit may provide important clues about general principles for cortical computation, extending well beyond the domain of space into the realm of thinking, planning, reflection, and imagination.

References

Buzsaki, G., and E. I. Moser. 2013. "Memory, Navigation, and Theta Rhythm in the Hippocampal-Entorhinal System." *Nature Neuroscience* 16: 130–38.

Couey, J. J., et al. 2013. "Recurrent Inhibitory Circuitry as a Mechanism for Grid Formation." *Nature Neuroscience* 16: 318–24.

Felleman, D. J., and D. C. van Essen. 1991. "Distributed Hierarchical Processing in the Primate Cerebral Cortex. *Cerebral Cortex* 1: 1–47.

Squire, L. R., and S. Zola-Morgan. 1991. "The Medial Temporal Lobe Memory System." *Science* 253: 1380–86.

Stensola, H., et al. 2012. "The Entorhinal Grid Map Is Discretized." *Nature* 492: 72–78.

RECORDING FROM MANY NEURONS SIMULTANEOUSLY

FROM MEASUREMENT TO MEANING

Krishna V. Shenoy

The human brain is comprised of approximately one hundred billion neurons, yet most of what is known comes from measuring the activity of one neuron at a time. Or, at the other extreme, studies rely on measuring the aggregate activity of thousands to millions of neurons at a time. This profound measurement limitation is changing rapidly. It is now possible to measure activity from many hundreds to thousands of individual neurons all at the same time, and it is widely believed that it will soon be possible to measure from many hundreds of thousands, or even millions, of neurons. As game changing as these breakthroughs are, several barriers to converting raw biological measurements into fundamental scientific meaning remain. Two of these challenges—making sense out of activity from large numbers of neurons and the importance of "levels of abstraction"—are discussed below.

Measuring Activity from Large Numbers of Neurons in the Brain

Neuroscientists seek to understand the function and dysfunction of the nervous system, including, ultimately, the human brain. The reasons for this pursuit are simple: to advance scientific knowledge about one of, if not the, most complicated systems in the universe as well as to help alleviate the burden of neurological disease and injury. In order to understand how a system like the brain operates one must measure its internal workings, much like understanding a computer requires measuring voltages and currents throughout its circuitry. In the case of the brain, this means measuring electrical activity (for example, action potentials, field potentials), chemical activity (for example, neurotransmitters, ion concentrations), and likely both throughout its neural circuitry. Pioneers in neuroscience have relied on various measurements in order to take accurate readings of electrical and chemical activity, and these measurements have tended to focus either on individual neurons (for example, intracellular electrode, extracellular electrode) or on aggregate activity from numerous neurons (for example, EEG, MEG, fMRI). Similarly, powerful stimulation technologies have been used to causally perturb neural activity and observe the consequences (for example, electrical microstimulation, TMS, optogenetics).

While many seminal discoveries, insights, and Nobel Prizes have resulted from these measurement (and stimulation) technologies, a renewed appreciation for the complexity of the overall nervous system and the associated need for measuring simultaneously from many individual neurons have arisen in recent years. Fortunately, technological innovation has risen to meet this need, making it now possible to measure from hundreds to thousands of individual neurons at the same time. For example, genetically encoded calcium indicators (for example, GCaMP, see chapter by Ahrens, this volume) allow calcium concentration changes associated with action potentials from thousands to tens of thousands of individual neurons to all be optically imaged simultaneously. The full potential of this class of measurement is still being realized, with animal models ranging from immobilized worms, walking transgenic mice, and freely moving rats already in use, to possibilities on the horizon including monkeys performing a variety of cognitive tasks.

More traditional electrode-based technologies have also scaled up in recent years. One example—a one-hundred-electrode array—is shown in figure 1a. One or more of these arrays can be implanted permanently in the brains of rats, monkeys, and humans (as part of FDA pilot clinical trials focused on neural prostheses to help people with paralysis). These electrodes can measure electrical activity (extracellular action potentials, field potentials) from tens to hundreds of individual neurons while animals perform a variety of cognitive tasks including sensory, decision making, and motor behaviors as shown in figure 1b.

a

b

fixate, touch   delay period   'go'   RT   movement   reach
                                 cue   onset

target
onset

0°

Neurons (ordered by
preferred direction)

360°

200 ms

c

Neuron 1
Neuron 2
Neuron 3

time

Spike trains

d

N_3

time

N_2

N_1

Noisy time series

e

N_3

S_2

N_1

S_1

N_2

Denoised time series

f

S_1

S_2

Low-dimensional
time series

Figure 1. Key steps in extracting low-dimensional, single-trial, neural population state-space trajectories from multichannel neural data. a. A silicon-based, 4 x 4 mm, 100 electrode array with 1 mm long electrodes made by Blackrock Microsystems Inc. The analysis begins with multineuron data from such a device. b. The instructed-delay center-out reaching task, with the neural response from an electrode array surgically implanted in premotor cortex. Central targets are fixated and touched with the eye and arm. A spot of light then appears on the screen indicating the target that should eventually be touched. Following a delay period a "go" cue is given, leading to an arm movement and touching of the target. "Brain states" associated with each part of this type of task can be measured from the activity of a population of neurons in the appropriate brain region(s). Each row of dots represents the times of action potentials (spikes) from one of 44 neurons. c–f. Computational steps for extracting a population neural trajectory from multiple spike trains on a single trial. For clarity, c illustrates spike trains recorded simultaneously from 3 (of the 44) neurons shown in panel b. d. Depicts the time evolution of the recorded neural activity plotted in a 3-dimensional space, where each axis measures the instantaneous firing rate of a neuron (e.g., N1 refers to neuron 1). Firing rates may be estimated in brief time bins (e.g., 10 ms). e. Depicts the population neural trajectory (a denoised version of the trajectory in d), which is shown to lie within a 2-dimensional space with coordinates S1 and S2. Finally, f. shows the population neural trajectory visualized directly in a low-dimensional state space and can be referred to using its low-dimensional coordinates (S1, S2). This final panel illustrates a low-dimensional, single-trial, neural population state-space trajectory computed from neural data measured simultaneously from many neurons. For further reading see Yu et al. (2009).

Even more revolutionary measurement technologies are also being developed, but the two technologies described above serve as examples that it is now possible to record simultaneously from hundreds to thousands of individual neurons.

Making Sense out of Activity from Large Numbers of Neurons

The first challenge to converting this newfound torrent of neural measurements into fundamental scientific meaning is to ask how to "make sense of the data." This is a deceptively simple-sounding question, as it would appear that we could just keep analyzing the measured data as we always have but now do so with a lot more presumably beneficial data. However, this would be overlooking many likely benefits of having massively parallel neural measurements where each neuron is measured with high temporal precision. Moreover, there may also be additional new information available such as cell type, axonal and dendritic projection pattern, and synaptic connection strengths. By analogy again to a computer, if presented with the opportunity to measure from one thousand transistors *simultaneously* it would save time relative to measuring one thousand transistors one at a time—but there are other far more important advantages as well. The reasons for this are developed more fully below.

Are there different ways forward? There are undoubtedly many potential ways forward, and at least one has been pursued in recent years and is termed the "dynamical systems approach" since it is borrowed and adapted from physical science and engineering where dynamical systems design and analysis is a staple. Three central elements to the dynamical systems approach are as follows. First, measured neural data constitute a time series, where there is correlation structure between measurements nearby in time. As such some form of temporal smoothing may be appropriate, and may help combat noise inherent in neural measurement. This is depicted in figure 1c–e. Second, the simultaneously measured neural data constitute a high-dimensional dataset but putatively actually occupies fewer dimensions. Dimensionality reduction, a major topic in machine learning and statistics, can be used to infer a lower-dimensional manifold on which the data reside. This is depicted in figure 1f. Taken together it is possible to visualize the nominally important dimensions that vary in the data and to then see how these population neural trajectories correspond to cognitive variables, such as the time it takes the arm to start moving following a "go" cue (reaction time, RT) or the direction in which the arm will move (see again figure 1b). It is important to note that very-low-dimensional visualizations, such as two- or three-dimensional figures drawn on paper, almost certainly miss some information. Thus such visualizations are useful for building intuition, but answering scientific questions must be done with higher-dimensional data where little if any information is lost. Finally, the dynamical systems approach seeks to estimate, quantitatively, the rules governing the evolution of the population neural state. This is akin to ascertaining Newton's laws from observations of a ball rolling on an uneven surface such that momentum, friction, and elasticity can be characterized. Together, visualizing lower-dimensional population neural trajectories, so as to generate hypotheses about how the neural circuit is working as a whole and relates to (single-trial) behavior, and identifying the equations of motion (for example, using a family of techniques known as systems identification) are a framework for leveraging massively parallel neural measurements into nominally meaningful scientific insights.

The Importance of "Levels of Abstraction"

The second challenge to fruitfully converting unprecedented volumes of neural data into scientific discoveries and insights—as opposed to potentially "drowning in data"—is to know what to pay attention to. This is certainly easier said than done when it comes to the brain, which is still poorly understood and it is unclear what details matter at a given level of investigation. Does the detailed connection pattern and synaptic strengths for each neuron matter when attempting to relate population neural activity to an arm movement? Does the exact pattern of action potential emission times matter when neurons must constantly contend with (probabilistic) synaptic failure? These questions, and countlessly many more, are open questions in neuroscience. Nevertheless, we can likely benefit by at least being aware that other fields in physical science and engineering contend with similar problems by adopting a well-proven philosophy for the design and analysis of physical systems.

This ubiquitous and essential concept to understanding and designing physical systems is termed "levels of abstraction." We anticipate that levels of abstraction will be of growing importance when investigating biological systems, including the brain. We describe here an analogy between a well-understood electronic system and the nervous system in order to highlight the potential merits of increasingly employing levels of abstraction in brain science.

Modern computer systems are comprised of several integrated circuits ("chips") connected together, and connected with peripherals such as displays, keyboards, and networked devices. Consider just one of these chips, the central processing unit (CPU), and how we can understand how it works. At the smallest level are atoms arranged precisely to bestow transistors with the desired electrical properties. Transistors come in a multitude of sizes and types, number in the billions, and form the next level of the CPU. The third level is the wiring between the transistors, which can be quite complex and have hundreds of millions of individual wires, due to wires bridging over and tunneling under each other similar to a metropolitan highway system. The fourth and final level, again broadly speaking, is the software. Software ranges from the detailed control of specific hardware (machine code) through the more global coordination of resources and data (operating system, algorithms). The software level is distinct from the other three because it resides in the pattern of electrical states (1s and 0s), as opposed to being physically manifest, and because it can grow to essentially arbitrary complexity by expanding well beyond the not uncommon millions of lines of code.

What does this have to do with the brain? Any detailed, literal comparison between the brain and a CPU is doomed. Examples of this type of flawed, detailed comparison that have been put forth in recent years include likening a computationally rich neuron to a computationally impoverished transistor (that is, a simple switch in a digital system), or likening the three-dimensional point-to-point connections between neurons to the essentially two-dimensional and relatively less general connections among transistors. Nevertheless, a broad comparison may help highlight how the levels of abstraction concept is anticipated to help shed insights on how the brain works. Importantly, this concept is related to David Marr's trilevel hypothesis, where in broad terms Marr's computational

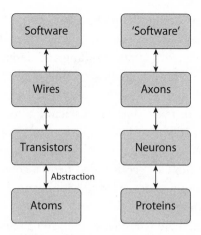

Figure 2. Levels of abstraction for a CPU (left column) and the brain (right column). Arrows indicate how detailed information at one level is abstracted away, so as to pass along only the essential operating principles and characteristics to the next level. Arrows are bi-directional to indicate that abstraction is beneficial both to understanding how physical implementation impacts software capability (bottom-up) as well as how software requirements impact physical design (top-down).

and algorithmic/representational levels are grouped, for brevity, into the software level, and his physical level appears here as the first three levels to reflect the increasingly detailed physical information available.

At the smallest level, there is similarity between material science focused on atomic design of silicon, dopants, and oxygen and molecular neuroscience focused on channel proteins, synapses, and neurotransmitters (see figure 2). While this detailed understanding is critical, some of the detail must be "abstracted away" in order to facilitate understanding (and the ability to design) at the next level, or else complexity will grow rapidly and the fundamental principles will be obscured. For example, aggregate properties and statistical descriptions of the materials must be brought forward, but specific locations of individual atoms must be left behind.

At the next level, there is similarity between device engineering focused on converting materials properties into transistors sizes and types so as to achieve the needed electrical properties and dynamics, and cellular neuroscience focused on neuron geometry, channel conductance, and membrane potential so as to understand electrical and neurotransmitter properties and dynamics. Again, the exquisitely interesting and important transistor designs must be abstracted away, passing to the next level only a few simple current-voltage rules. Without such abstraction, or simplification, understanding and designing the next level would be intractable both analytically and computationally. What to include or exclude when abstracting away detail when it comes to cellular neuroscience, or molecular neuroscience before that, is of course an open question, and we do not propose an answer. Instead we highlight the need for this question to be addressed, since for many physical systems, including the CPU considered here, a comprehensive understanding and the ability to design would simply not be possible without abstraction between levels.

At the third level, there is similarity between circuit design and computer architecture focused on the optimal wiring between transistors and between chips, and neuro-anatomy and connectomics (see chapters by Sporns, Zador, and Hawrylycz, this volume) focused on the detailed wiring and wiring rules between neurons within a brain area and between brain regions. Again abstraction is essential in the CPU case as the overall hardware capabilities and limits are of paramount importance when working at the next (software) level, and, similarly, it is anticipated that the overall neural "hardware" capabilities and limits are of primary importance when working at the next (neural "software") level. How best to abstract away detail in the neural context is again an open question, perhaps especially so as the neural hardware changes through time (that is, development, learning, plasticity), unlike most electronic hardware.

At the fourth and final level, there is similarity between computer architecture and computer science—focused on designing machine codes, operating systems, and algorithms that orchestrate all information processing—and on systems and cognitive neuroscience, including network modeling—focused on the relationship between neural activity and sensation, perception, decisions, actions, and more abstract thought. In broad terms, this is the level of the CPU that faces the greatest challenge if the levels of abstraction discipline is not followed. This is because inheriting the full complement of details from the three prior levels would leave one attempting to understand an existing CPU (that is, reverse engineering) or designing a new CPU hopelessly confused in the morass of information; without any prioritization as to the properties that are of direct relevance and those that, while critical to each prior level, are no longer essential to understanding at the final level one cannot see the forest through the trees.

With the levels of abstraction concept in place, it becomes possible to glean new insights into the fundamental operation of a CPU at this final level and, we anticipate, the same will be possible for the brain. As an example, consider what could be learned about a CPU with a few hundred oscilloscopes. With one oscilloscope it is possible to measure the electronic waveform from one transistor terminal, discover that voltages tend to be either high or low (that is, binary), see that voltages change very fast (for example, ns) and do so according to a master clock (for example, 5 GHz), and one could then conjecture that the transistor is part of an adder, memory register, or data bus. Moreover, if it is possible to place the CPU in exactly the same state again and now measure from a different transistor terminal it should be possible to, across many such measurements, build up a more complete picture.

If instead a few hundred oscilloscopes measure a few hundred transistor terminals at the same time then it is possible to discover additional crucial properties of the CPU. This includes how transistor states are coordinated through time (that is, circuit dynamics), how the system functions during normal operation where the same exact set of transistor states may seldom if ever be seen twice, and to postulate the essential features of the software. For example, it is possible to understand the fine-timing coordination principles among a set of transistors responsible for adding two numbers, as well as to understand how faulty coordination between transistors (that is, a timing "glitch" caused by a design "bug") leads to arithmetic mistakes, all without needing to have the same two numbers added repeatedly and all control circuitry in precisely the same state, which may be essentially impossible. This is possible by virtue of simultaneous measurements, dimensionality reduction and dynamical systems analysis methods and modeling, and, again, levels of abstraction—which assures that detailed knowledge of atoms, transistor sizes/types, and wiring that are not essential to proceed with analyses well suited for this final level of investigation do not cloud the investigation or answers. Similarly, we anticipate that measuring from hundreds to thousands of neurons simultaneously and analyzing these data with methods capable of revealing fundamental operating principles (for example, dimensionality reduction, dynamical systems, network modeling) should now be possible and insightful. For example, it may now be possible to understand how populations of neurons in the brain make decisions based on a constant, and seldom if ever repeated, flow of sensory and goal information experienced as part of everyday life.

It is important to note that while, for simplicity, the four broad levels are described from "bottom up" and the importance of levels of abstraction is also emphasized in this unidirectional fashion, this is only half the story (see figure 2). In the CPU analogy is it equally important to apply levels of abstraction starting at the fourth level (for example, what general classes of software/algorithms need to be supported) and proceeding toward the first level (for example, what materials are needed to support a certain type of transistor performance). This also completes the design cycle, as well as moves closer to a comprehensive understanding, by relating the software/system requirements all the way to the materials and transistor choices and tradeoffs. One would expect this to also be the case with neural systems. A better understanding of the key neural computational principles should help deepen understanding of anatomical connection patterns, single neuron computation, molecular underpinnings and their various design trade-offs.

Summary

We are currently in the midst of a neurotechnology revolution that is making it possible to measure (and stimulate) thousands and potentially millions of neurons simultaneously. This unprecedented access to neural data is on the one hand extremely exciting and on the other hand profoundly humbling. What will we do with all of these data? How will we make sense out of it all, and how can we even begin to think about what details matter to each level of understanding and question being posed? While it is tempting to carry on with inherently single-neuron-oriented analyses, or to treat this unique neural dataset as just another "big data" dataset and unleash somewhat generic machine learning algorithms on it, both would likely limit the full extent of insights that are believed to be possible. We discussed here just two of the key challenges moving forward, and we offer two possible approaches—dynamical systems analyses and the levels of abstraction philosophy.

Further Reading

Churchland, M. M., B. M. Yu, M. Sahani, and K. V. Shenoy. 2007. "Techniques for Extracting Single-Trial Activity Patterns from Large-Scale Neural Recordings." *Current Opinion in Neurobiology*, special issue on new technologies (17): 609–18.

Diester, I., M. T. Kaufman, M. Mogri, R. Pashaie, W. Goo, O. Yizhar, C. Ramakrishnan, K. Deisseroth, and K. V. Shenoy. 2011. "An Optogenetic Toolbox Designed for Primates." *Nature Neuroscience* (14): 387–97.

Gilja, V., P. Nuyujukian, C. A. Chestek, J. P. Cunningham, B. M. Yu, J. M. Fan, M. M. Churchland, M. T. Kaufman, J. C. Kao, S. I. Ryu, and K. V. Shenoy. 2012. "A High-Performance Neural Prosthesis Enabled by Control Algorithm Design." *Nature Neuroscience* (15): 1752–57.

Shenoy, K. V., M. Sahani, and M. M. Churchland. 2013. "Cortical Control of Arm Movements: A Dynamical Systems Perspective." *Annual Review of Neuroscience* (36): 337–59.

Yu, B. M., J. P. Cunningham, G. Santhanam, S. I. Ryu, K. V. Shenoy, and M. Sahani. 2009. "Gaussian-Process Factor Analysis for Low-Dimensional Single-Trial Analysis of Neural Population Activity." *Journal of Neurophysiology* (102): 614–35.

NETWORK NEUROSCIENCE

Olaf Sporns

Unraveling the mechanisms and principles that create mind and cognition from the activity of roughly eighty-six million neurons in the human brain remains one of the most alluring as well as urgent scientific pursuits. The urgency of the task is underscored by the immense and growing health, social, and economic impact of brain and mental disorders. How can we get closer to a more complete understanding of how the brain works? Certainly, progress will in part come from the slow and gradual accumulation of increasingly more detailed insights about neural mechanisms. But as I will argue in this essay, neuroscience will also need to shift perspective, toward embracing a view that squarely acknowledges the brain as a *complex networked system*, with many levels of organization, from cells to cognition, that are individually irreducible and mutually interconnected.

Indeed, connectivity is a core theme of the age we live in. For humans, connectivity (to the Internet that is) is now considered to be so fundamentally empowering that the United Nations has proposed it as a basic human right. For scientists across a wide range of disciplines, connectivity is emerging as a major focus for understanding and managing the behavior of complex systems. At the forefront of these efforts, biological researchers have grown increasingly aware of complex networks formed by interactions among molecules and cells, giving rise to the entirely new field of systems biology. As it turns out, the functioning of cells critically depends on gene regulatory, signaling, and metabolic networks that shape interactions among molecules, and interactions among cells are crucial for building and maintaining whole organisms.

This core theme of connectivity is now rapidly gaining ground in neuroscience. There is a growing realization that virtually all aspects of integrative brain function depend on the action of networks—those created by connections among neurons and brain regions. These connections are vitally important for neuronal information processing and computation, and disturbances of connectivity appear to be associated with most disease states of the brain. The important role of connectivity in brain function is the main motivation behind the drive to create a complete map of the brain's connections, now commonly called the connectome (see chapter by Zador, this volume). The main thesis of this essay is that the current and future growth of *network neuroscience*, the emerging science of brain networks, will fundamentally advance our understanding of brain function.

Moving from Circuits to Networks

Of all organs of the human body, the brain's function is perhaps the hardest to succinctly define. Simplistic and overly reductive notions tend to run into serious problems. To mention a colorful example, the nineteenth-century German scientist Karl Vogt once wrote that "thoughts stand in the same relation to the brain as gall does to the liver or urine to the kidneys." When he expressed this idea in public, a philosopher interjected that the longer one listens to Professor Vogt, the more one tends to believe him. Clearly, more sophisticated ideas and models are in demand.

One reason for the difficulty in understanding the brain is that there is a vast gap between the functioning of any one of its billions of nerve cells and the functioning of the brain as a whole. Looking back on the history of neuroscience, some of the most fundamental insights have unquestionably come from studies of single neurons. Indeed, the "neuron doctrine," the notion that the nerve cell or neuron is the fundamental unit of brain function, has been the unshakable foundation of modern neuroscience for over a century. Thousands of studies employing a broad range of tools from microelectrode recordings to functional neuroimaging have shown that neurons can exhibit often remarkably specific response properties. The activity of individual neurons can represent highly complex stimuli or events such as the appearance of a person's face, the sound of a familiar voice, the contraction of eye muscles, or the direction of limb movements. Going beyond responses to current inputs, activity in specific neurons is involved in recalling past experience or in signaling the anticipation of future reward. Neuronal activity is associated with virtually the entire repertoire of complex mental processes like working memory, attention, visual imagery, or even dream states.

What endows neurons with these remarkably diverse response properties? The answer is (at least) threefold. First, neurons express extremely complex sets of molecules that generate and sense the electrical signals that underlie all neuronal responses and synaptic transmission. Second, neurons have very distinct shapes or morphologies that play an important role in the way synaptic inputs are converted into the neuron's output, usually in the form of an action potential. But molecules and morphology are not enough to explain neuronal responses—a major role is played by neuronal *connections*, the synaptic links that tie neurons together into vast networks. Take away a neuron's connections and it becomes deaf and mute, cut it off from inputs and it becomes unable to exert any influence whatsoever. The power of neurons derives from their collective action as part of brain networks, bound together by connections that allow them to interact, compete, and cooperate. "Brain cells fire in patterns," as Steven Pinker once put it when challenged on the Colbert Report to explain brain function in five words. And these patterns are orchestrated by connections.

Although we have known for a long time that neurons are connected into circuits, and that it is this circuit activity that drives all perception, thought, and action, I would argue that modern concepts of networks add an important new dimension. The more traditional way of thinking in terms of circuits is based on the notion of highly specific point-to-point interaction among circuit elements with each link transmitting very specific information, much like an electronic or logic circuit in a computer. The action of the circuit as a whole is fully determined by the sum total of these specific interactions. A corollary is that circuit function is fully decomposable (given complete data) into neat sequences of causes and effects. In this sense, circuits resemble Laplacian models of classical mechanics, with circuit elements exerting purely local effects on each other and with connections mediating specific causal roles.

In contrast, modern approaches from complexity theory and network science emphasize that global outcomes are irreducible to simple localized causes, and that the functioning of the network *as a whole* transcends the functioning of each of its individual elements. One key concept is that of "emergence." Emergence builds on the basic observation that collective interactions among the elements of complex networked systems often give rise to new properties that do not exist at lower levels of organization. In the case of large networks of neurons, powerful manifestations of "emergent phenomena" are global states of brain dynamics in which large populations of neurons engage in coherent and collective behavior. Such dynamics emerge from very large numbers of local interactions that are individually weak and yet collectively powerful enough to create large-scale patterns. Take as an example the phenomenon of neural synchronization, the coordinated firing "in sync" of large numbers of nerve cells. Synchronization clearly depends on neuronal interactions mediated by synaptic connections, but it is not attributable to any specific causal chain of interactions in a circuit model. Instead, synchronization is the global outcome of many local events orchestrated by the network as a whole, the coordination of a myriad of weak dynamic couplings along connections between nerve cells.

Synchronization is only one example of global network dynamics that results from many elementary network interactions. Another particularly intriguing example is the so-called critical state, a dynamic regime where a system engages in a wide range of flexible and variable behaviors. Poised between order and randomness, the critical state may allow neuronal systems to exhibit a rich repertoire of dynamic patterns, balance sensitivity to current input with memory of the past, and show a high capacity for computation. While criticality can occur in many different types of networks, it does appear that some network attributes, including several that have been found in brain networks, promote or stabilize critical dynamics. While much work remains to be done in drawing links between network architecture and brain dynamics, shedding light on how networks shape cooperative interactions among interconnected neurons will likely add a new dimension to more traditional views of neuronal circuits.

Meeting the Challenge of "Big Data"

Collective and emergent behavior of complex networks is ubiquitous. It is found not only in the brain but also in other biological systems like cells and ecosystems, and even in social and technological networks.

The effects of such collective network dynamics are all around us. In fact, we are personally becoming more and more enmeshed in systems whose emergent behavior can be very nontrivial and hard to predict, with sometimes far-reaching consequences for well-being and survival (think financial systems and climate change). Significant scientific efforts are underway to more accurately model how the dynamics of sociotechnological systems can impact economic stability, influence the spread of global pandemics, or trigger the onset of revolution or war. These efforts are fueled by an ever-increasing ability to record, store, and mine digital data on social and economic behavior. The advance of what is currently fashionably called "big data" appears unstoppable.

Neuroscience, it turns out, is on the brink of its own "big data" revolution. Supplementing more traditional practices of small-scale hypothesis-driven laboratory research, a growing number of large-scale brain data collection and data aggregation ventures are now underway, and prospects are that this trend will only grow in future years. For example, the European Community has just embarked on a ten-year quest to simulate a human brain in a supercomputer (see chapter by Hill, this volume). In the United States, the Human Connectome Project, a comprehensive survey of human brain connectivity across the healthy adult population will soon (by 2015) yield more than a petabyte of high-quality brain imaging data. Even more ambitious proposals to create accurate maps of how neurons connect at the synaptic scale would result in even more outlandish amounts of data. Mapping the roughly one hundred trillion synaptic connections of a human brain would by some estimates generate on the order of a zettabyte of data (a zettabyte corresponds to one million petabytes). For comparison, that's about equal to the amount of digital information created by all humans worldwide in the year 2010. And recently proposed efforts to map the human brain's functional activity at the resolution of individual cells and synapses might dwarf even these numbers by orders of magnitude. The coming data deluge is likely to transform neuroscience, from the slow and painstaking accumulation of results gathered in small experimental studies to a discipline more like nuclear physics or astronomy, with giant amounts of data pulled in by specialized facilities or "brain observatories" that are the equivalent of particle colliders and space telescopes.

What to do with all this data? Physics and astronomy can draw on a rich and (mostly) solid foundation of theories and natural laws that can bring order to the maelstrom of empirical data. Theories enable significant data reduction by identifying important variables to track and thus distilling a torrent of primary data pulled in by sophisticated instruments into interpretable form. Theory translates "big data" into "small data." A remarkable example was the astronomer Edwin Hubble's discovery of the expansion of the universe in 1929. Integrating over years of observation, Hubble reported a proportional relationship between redshifts in the spectra of galaxies (interpreted as their recession speeds) and their physical distances. Viewed in the context of cosmological models Albert Einstein and Willem de Sitter formulated earlier, these data strongly supported cosmic expansion. This monumental insight came from a dataset that comprised less than fifty data points, compressible to a fraction of a kilobyte. When it comes to applying theory to "big data," neuroscience, to put it mildly, has some catching up to do. Sure enough, there are many ways of analyzing brain data that are useful and productive for extracting regularities from neural recordings, filtering signal from noise, deciphering neural codes, identifying coherent neuronal populations, and so forth. But data analysis isn't theory (see also chapters by Freeman and Shenoy, this volume). At the time of this writing, neuroscience still largely lacks organizing principles or a theoretical framework for converting brain data into fundamental knowledge and understanding.

Network science may be one appealing candidate for offering such a theoretical framework. Network approaches have already proven extremely useful for organizing and interpreting big brain data. One area where network concepts are sharply on the rise is cognitive neuroscience, especially studies that use noninvasive neuroimaging to map human brain activity. Traditionally, there was much interest in isolating specific regions of the brain that became activated in association with specific stimuli, mental states, or tasks. More recently, there has been a shift in interest from activation to *coactivation* studies that take into account not only which regions are active but also their dynamic interactions that result in networked brain activity. This shift was catalyzed by the realization that the brain is never completely inactive, even when a person is awake but does not engage in externally cued attention-demanding cognitive processes, a rather unconstrained and task-free state often referred to as "rest." The resting brain turns out to be a cauldron of activity, which is both seemingly spontaneous and highly organized into spatial and temporal patterns. Many aspects of these patterns are shared across individuals and, at least in part, reflect the anatomical connections that link brain regions to each other. Application of network science tools and methods have revealed numerous "resting-state networks," sets of brain regions whose activity is highly correlated in the course of resting brain activity. Importantly, these resting-state networks closely resemble sets of brain regions that are consistently coactivated as the brain is challenged across a broad range of sensory inputs or tasks. For example, tasks that require the direction of attention to a portion of the external environment, such as quickly detecting targets in cued locations of the visual field, reliably activate a distributed set of specific brain regions in the frontal and parietal cortex. At rest, the same set of regions is found to undergo correlated fluctuations in neural activity. These relations between task-driven and task-free patterns of activity are consistent with the idea that the resting brain rehearses or recapitulates a set of network states, each of which are associated with different domains of human cognition.

As these studies at the large scale of whole brain activity suggest, distributed networks rather than localized brain regions may be fundamental units of how the brain is organized and how it responds. Network approaches have not only been instrumental in revealing this organization, they are also increasingly important for tracking down biological substrates of brain and mental disorders. Specific disturbances of how brain networks are configured and how they dynamically respond and interact have been documented in a range of disorders, from neurodegenerative conditions such as Alzheimer's disease to mental illnesses such as schizophrenia. A common theme is that of disconnection—a disturbance or loss of connectivity among specific neurons and brain regions that manifests in specific impairments of integrative cognitive ability. For example, in the case of schizophrenia numerous studies comparing connectivity in the brains of patients and healthy controls have shown an association of clinical symptoms with impaired functional coupling between parietal and prefrontal regions, possibly the result of "mis-wiring" in long-distance interregional projections. Such mis-wiring not only disrupts specific pathways and connections, it also results in global changes in the way the entire network processes information—somewhat like the widespread disruption of traffic patterns after closure of a single highway or air transportation hub.

Going beyond diagnosis, network approaches may also become important for developing new interventions and therapies. If network disturbances underlie common disease states, effective therapy and recovery may involve coaxing the disrupted network back into a regime where functionality is restored. A surprising adjunct in this endeavor of finding ways to treat brain networks is the use of sophisticated computer models that can reproduce and predict the dynamic activity of human brain networks.

Building Virtual Brains

Brain models have evolved tremendously over the past two decades, from the historically useful but biologically unrealistic constructs of "artificial neural networks" to biologically based computational models that combine the major ingredients of neuronal biophysics and connectivity to create realistic brain dynamics (see chapters by Hill and Eliasmith, this volume). These models are computational analogues of complex brain networks, and they are beginning to provide fundamentally new insights into how brains respond and compute. As models, their construction is simple—sets of neural elements and their anatomical interconnections, the latter typically derived from empirical measurements, and a set of dynamic equations that are based on the electrical response properties of nerve cells. The complete model is set in motion by exposing it to external inputs as well as some source of internal noise. Once in motion, the model's neural elements produce simulated activity traces that can be analyzed and processed in ways that closely resemble how scientists look at experimental data. A big advantage of the modeling approach is that unlike the empirical brain, the model's internal workings are completely known and the model's structure can be modified in order to explore how its activity changes. Models of this kind have already provided important insights. Think back to the "resting" brain—given that there are no explicit tasks or inputs, what accounts for the reproducible spatiotemporal patterns that are so characteristic of resting brain activity? Computational models have shown that these patterns strongly depend on a combination of several factors, including the layout of structural connections among brain regions, balanced local coupling of excitatory and inhibitory neuronal populations, and the presence of conduction delays and dynamic noise. Take away any of these ingredients and the activity of the model will no longer match what is empirically observed.

Computational models have become powerful tools in many disciplines, from astrophysics to traffic engineering, and they will play an increasingly prominent and indispensable role in neuroscience. Of particular importance will be network-based models of "virtual brains"—models that, not unlike the global simulators employed in climate forecasting, allow drawing links between variations in local parameters that determine brain connectivity and resulting changes at the global scale, for example, those manifesting in patterns of brain dynamics. In the near term, such models will become computational platforms to explore the effects of localized brain lesions such as those that might result from stroke or brain trauma on network communication across the remaining brain. In the middle term, more sophisticated models will begin to implement models of pathophysiological processes that can mimic the progression of disease states involving neurodegeneration or developmental abnormalities. In the longer term, computational models based on individual patient data may become useful tools for designing therapeutic interventions that are tailored to that patient's very own brain network—perhaps opening the door to "personal connectomics" as a component of clinical practice.

These developments are perhaps closer than most of us think. One reason is the continued rapid rise in computational power per unit cost. Another important reason is the convergence of network science approaches across a broad expanse of biological, social, and technological systems. This convergence creates enormous opportunities for synergy and collaboration that would have been unthinkable even a few years ago. It will also drive the development of new recording probes and observational tools. The growing realization that brain function depends on connectivity and network interactions among many elements and processes mandates the development of sophisticated empirical and analytic methods for mapping and tracking these network interactions. Finally, linking brain networks to behavior will mean stepping beyond the boundaries of the nervous system to consider how connectivity within the brain is modulated by the dynamic coupling of brain and environment (see chapter by Ahrens, this volume). Neurons don't just passively respond to inputs—by contributing to motor activity and behavior they actively determine what the inputs are. Capturing this dynamic interplay between brain and behavior will require an extension of the concept of functional connectivity from intrinsic brain networks into the environment. Clearly, the enormous complexity of brain networks will pose formidable challenges for the foreseeable future.

As I look to the future, it seems inevitable that neuroscience will continue to move from focusing on components to mapping and modeling their interactions, building on a reconceptualization of the brain as a complex networked system. I expect that this shift towards *network neuroscience* will lead to fundamentally new insights. As many studies have shown, the organization and architecture of networks from a surprising range of real-world systems (cells to society) express a set of shared and common themes and motifs. Network neuroscience suggests that the brain is another example of such a system. So, perhaps the brain is less special than we previously thought. While the brain is certainly unique in that it mediates all personal experience, we may find it does so by following a set of general and universal laws that govern the functioning of complex networks.

Further Reading

Sporns, O. 2011. *Networks of the Brain*. Cambridge, MA: MIT Press.

Strogatz, S. 2004. *Sync: How Order Emerges from Chaos in the Universe, Nature, and Daily Life*. New York: Hyperion.

Swanson, L.W. 2011. *Brain Architecture: Understanding the Basic Plan*. New York: Oxford University Press.

LARGE-SCALE NEUROSCIENCE

FROM ANALYTICS TO INSIGHT

Jeremy Freeman

The brain contains millions of neurons. Equally vast is the range of any creature's experience in an ever-changing world.

Yet the laboratory typically limits both the scale of neural measurement and the complexity of behavioral context. An experimenter might record the response of a single sensory neuron to a tiny set of stimuli in order to determine which stimulus triggers the most vigorous response. Or an experimenter might measure the activity of a few isolated motor neurons while an animal performs one simple behavior, which allows the researcher to establish a clear relationship between the behavior and the neural responses. By such simplification, however, we might be missing the forest for a handful of trees. Almost surely, a complete understanding of the brain will demand a more holistic approach: complex behavior reflects information processing across the entire nervous system, involving the coordinated activity of thousands or millions of neurons of diverse type and function within and across multiple circuits and brain areas. New technologies are now, finally, allowing us to probe the activity of thousands of neurons simultaneously while animals perform rich, ethologically relevant behaviors: mice running on balls exploring virtual mazes, fish swimming against moving backgrounds, or flies flying toward virtual targets.

But what do we do now that we have these tools? How do we begin to understand the vast quantities of data we are beginning to collect? Standard techniques can help with some of the basics, like extracting signals from noise—if we know where to look, and if we can handle the scale of the data. But how do we use the mountains of data to extract a rich theoretical understanding of how the brain really works?

$$\bullet \quad \bullet \quad \bullet \quad \bullet \quad \bullet \quad \bullet$$

The first thing to understand about neural data is that the immediate output of an experiment—for example, a set of images of calcium fluorescence from a microscope—is never a straightforward list of neural responses at each moment in time. Rather, we always have to *infer* what the brain is doing because the images themselves are messy and indirect. Images of neural activity are rife with movement and other artifacts, akin to the grainy, often blurry photos taken by early digital cameras. As a result, the first step in neural data analysis is almost always to address the low-level challenge of sorting signal from noise. Computer algorithms can process source signals, remove artifacts, and extract the relevant components. In some cases the goal of these initial steps is relatively straightforward, for example, minimizing the motion between successive frames. But finding the best solution still poses a challenge. In other cases there is a surprising degree of subtlety. For example, in each of the images from a calcium imaging experiment there are millions of pixels. Depending on the neural circuit, cell type, and calcium indicator under study, those pixels correspond to a complex mixture of cell bodies and extensions from those cell bodies, including axons and dendrites. To analyze the data, we can use sophisticated algorithms to isolate the signals from cell bodies, discarding the rest of the image; alternatively, we can interrogate the response of every single pixel, capturing as much functionally relevant information as possible, but losing the landmarks that cell bodies provide. In truth, there is no absolute answer; the only consensus thus far is that the choice will probably depend on the scientific question asked. Even at this early stage of analysis, computer algorithms play a critical role in data analysis. And good scientific judgment is key in sorting through the many options.

Higher levels of analysis aim to find patterns in neural responses that are related to sensory input, behavioral output (and concomitant sensory feedback), behavioral state (as induced by modulatory processes), or most likely, to a complex mixture of all three. The biggest challenge is never knowing ahead of time the "right" analysis. Consider measuring neural responses during a fairly straightforward task: a mouse runs on a ball, is presented with a few different visual stimuli on a screen, and then must indicate a response. Even here, in a situation far from the rich experience of the mouse roaming in the real world, many parameters are potentially relevant to neural responses. Those include the stimulus itself, how fast the animal was running, what her behavioral response was, whether she got the answer right or wrong, and aspects of her state that might be difficult to characterize explicitly, such as attention or arousal.

How do we relate neural responses to all of these parameters? How are the parameters in turn related to one another? How should features of the behavior be represented for such analysis? Answering these questions falls under the broad category of "functional modeling," which aims to relate neural responses to observable features of the external world (see chapter by Carandini herein). Of course, the goal is not simply to reproduce the complete neural response pattern, like a high-tech tape recorder; the goal instead is to characterize what the brain is doing in terms that can aid our intuition and understanding and help consolidate our hypotheses. For example, the various neurons within a circuit, due to their type, morphology, projections, or laminar specificity, may differ in the degree to which their responses reflect each of the properties listed above. Functional modeling can begin to paint a picture of the computations these different elements of the circuit are performing. Coupled to appropriate anatomical information, they may reveal a picture of what the circuit does, what it is for.

When studying more complex representations, like those involved in motor control, the joint dynamics of neurons are particularly critical, and examining the response of one neuron at a time—or the grand average of all neurons—yields an incomplete, and potentially misleading, picture. A set of relatively theory-neutral techniques known as dimensionality reduction can sometimes uncover simpler structure hidden in high-dimensional patterns of joint activity (see chapter by Shenoy). But a tacit assumption of these methods is that all neurons under study are of the same kind and are doing roughly the same sort of work. This assumption breaks down when we examine larger fractions of the brain simultaneously and consider the extraordinary diversity and specificity of neural circuits. What if neurons of a particular type only respond when a pattern of response is present in another cell type, as might arise when a circuit's function is switched on or off by modulatory processes? What about "command" neurons that may be few in number, respond only once in a long experiment, and thus remain hidden from low-dimensional representations but are in fact crucial to the circuit's computation? In such cases, we might not find such features in the data unless we deliberately look for them. Prior theoretical principles or hypotheses—as opposed to pure bottom-up data mining—can become part of the analysis process. Sometimes it may be enough to "let the data speak for itself," but even that "speaking" can proceed only through active and creative data exploration, not brute force.

In order to perform flexible and exploratory analytics on neural data, we must scale up our analyses to handle the massive datasets we are starting to collect. A typical two-photon calcium imaging experiment that monitors responses over a small region of mouse cortex (~1,000 neurons) yields 50–100 GB of data, and a typical light-sheet imaging experiment that monitors responses over the entire brain of the larval zebrafish (~100,000 neurons) currently yields 1–5 TB (see chapter by Ahrens). Faster frame rates and longer experiments could soon yield 100–200 TB per experiment. By comparison, the social networks Twitter and Facebook collect hundreds of TBs of data from their users every day. Neuroscience is thus quickly entering the realm of massive, web-scale data.

At these scales, even simple analyses can take hours or days to perform, and more complex analyses are sometimes out of the question. Exploratory data analysis requires trying many analyses, and if each one takes a day, the size of the data becomes a major bottleneck for progress. Fortunately, the massive investments in "big data" from places like Google and Amazon have yielded novel approaches, like Google's MapReduce and the open-source counterpart Hadoop, which enable large-scale distributed computing; a terabyte can be analyzed by a cluster distributed through the cloud, rather than on a single machine at an individual investigator's desktop. An automatic front end to a large network of computers allows a scientist to focus on the goals of her algorithm rather than on the details of how work is distributed, scheduled, and executed across the cloud, thus enabling new analyses that would otherwise be unthinkable.

In my own lab, we have been adapting one of the most recent, and most exciting, large-scale data processing platforms—an open-source project called Apache Spark—to the problem of neural data analysis. Although primarily used in industry, Spark is ideally suited to the challenges of neuroscience data analysis because of three key advances. First, it introduces a primitive for data sharing that allows distributed data to be cached (that is, preloaded). When analyzing neural data, we often want to load it, apply all the kinds of analyses described above, and interactively inspect the results. Rather than reload the data for every analysis, Spark can cache TBs of data into distributed RAM, enabling rapid access as though the data were on a local machine, as well as supporting complex algorithms involving many iterations. Second, Spark provides powerful abstractions, accessible through its APIs (application programming interfaces) in Python, Java, and Scala, that make it easy to write and prototype analyses, helping the user focus on the algorithm, not the details of implementation. Third, Spark includes a module, Spark Streaming, for doing real-time analyses on streaming data. Spark itself is a set of primitives for distributed computing, and we have built a library on top of it, called Thunder (http://github.com/freeman-lab /thunder), for fast, easy, and interactive large-scale neural data analysis and visualization. With this library, analyses that would have previously taken a day can run within seconds or minutes, including regression, spatial and temporal factorization, time series modeling, and ever more complex alternatives. This library provides us, and the neuroscience community, with a nearly unprecedented opportunity for exploratory, large-scale neural analytics.

• • • • • •

Scaling up analysis is crucial, but it is just a prerequisite for progress, not progress itself. Especially when studying more complex behaviors, or more complex aspects of sensory representation, nearly an infinite number of inputs could be presented to an animal, or behaviors monitored, or task configurations tested. Effectively exploring these options is not a problem of data analysis but rather of designing targeted experiments in lockstep with the analysis. In many cases I expect the final answer will come not from an analysis but from a clever, and surprisingly simple, experiment.

As an example, I will describe a story from my graduate work in the visual system of the primate. The work concerned how visual signals are represented and transformed along the processing pathway of the primate visual system. It has long been known that the earliest stages of vision—the retina, and the thalamus—represent the presence of light and dark across our field of view. The first stage of cortical processing, primary visual cortex (or V1), represents not only the presence of light but also its shape. The responses of neurons in V1 specifically depend on whether a pattern, like a bar or an edge, is oriented vertically, horizontally, or diagonally. The discovery of this "orientation selectivity" by David Hubel and Torsten Wiesel in 1959 provided a fundamental insight into the cortical representation of visual information. But there exists a hierarchy of areas beyond V1, including V2, V3, V4, and several more, each of which are thought to represent increasingly complex, and behaviorally relevant, aspects of a visual scene. Our understanding of most of these areas remains limited, but V2 has been one of the most perplexing. Higher areas have been found to contain neurons selective for complex objects and shapes, like faces, but despite decades of work, the "function" of V2 neurons has remained mysterious.

One approach to characterizing visual responses is brute-force data collection: we measure responses to a large ensemble of random patterns of light and dark and examine the subset of patterns that elicited a response to reveal what the neuron encodes. This approach succeeds in V1 (revealing oriented edges), but it fares poorly in V2. We speculated at the time, and now know with some confidence, that this is because responses in V2 selectively encode visual elements that are significantly more complex than those encoded in V1 and thus are extremely unlikely to arise in a random pattern, no matter how long the experiment. They are also difficult to identify through intuition (unlike, say, faces or objects).

As an alternative to brute-force data collection, we developed a hypothesis of what V2 might be doing by integrating several pieces of computational and neuroscientific knowledge: the kinds of inputs V2 neurons received (predominantly from V1), what image properties could theoretically be represented by performing computations on those inputs, and which of those properties are found in the images encountered by an animal and would thus be behaviorally useful to encode. The key of the hypothesis: V2 neurons care not only about the presence of oriented patterns but also about their organization and statistical codependencies. In other words, they care whether, for example, patterns of two orientations or sizes tend to occur alongside each other in an image. We used this hypothesis, coupled to a computer algorithm for image synthesis, to construct targeted visual stimuli to present to V2 neurons. We then performed experiments in which we recorded responses of neurons in both V1 and V2 to these stimuli and found that responses to our stimuli differentiated the two areas better than in any previous reports.

These experiments have not, as yet, fully explained the function of V2. They were also fundamentally limited: we recorded the activity of a small number of neurons in each area, with no knowledge of cell type or laminar specificity, in the context of a highly unnatural behavior (an animal looking at images on a screen). But as we acquire richer and more complex tools and datasets in other systems, the moral is worth remembering: we arrived at an important insight about neural computation not through the collection of a massive dataset, or even a complex analysis, but rather through computationally driven insight and carefully designed experimental stimuli.

The era of large-scale neural data analysis is just beginning, but the importance of analytical and experimental interplay is already at the core of my lab's new work in this domain. In a collaboration with Misha Ahrens, for example, we are using whole-brain calcium imaging in the larval zebrafish to monitor responses while animals perform tasks in which simple sensory inputs elicit motor behaviors. We develop large-scale analyses that try to identify and disentangle signals related to sensory processing, motor behavior, ongoing neural dynamics, or some complex mixture of all three; and these analyses provide tantalizing hints as to the spatiotemporal structure of neural computation. But the same analyses often motivate new experiments in which we isolate each of these different components through controlled, yet still ethologically relevant, behaviors. Making sense of these data in turn requires new analyses, although often more targeted than the exploratory analyses that motivated the experiment. Analysis and experiment remain in constant interplay.

This moral is crucial as we prepare neuroscience for the future. A presumption emerging in our field is that the strategy for success is to collect masses of data, and then, only afterward, to distill from that data an understanding of how the brain works. In some domains, this rather static strategy—collect data first, analyze later—may be both reasonable and profitable. Take, for example, the problem of segmenting neurons from anatomical images to identify connectivity. Achieving that goal will demand powerful algorithms, but the goal itself is clear, so the analysis can proceed somewhat independently of data acquisition and experiment. But the more we stray from such well-defined problems, the less realistic that sort of static strategy may be. In most cases, we do not quite yet know which data we want to collect. Even if it is clear which kinds of measurements we want to make (for example, whole-brain calcium imaging of the larval zebrafish, two-photon imaging of multiple areas of mouse cortex), it is not clear which behaviors the organism should be performing while we collect those data, or which environment it should be experiencing. It is hard to imagine a single dataset, however massive, from which the truths we seek will emerge with only the right analysis, especially when we consider the nearly infinite set of alternative experiments we might have performed. Instead, we need an iterative process by which we move back and forth between using analytic tools to identify patterns in data and using the recovered patterns to inform and guide the next set of experiments. After many iterations, the patterns we identify may coalesce into rules and themes, perhaps even themes that extend across different systems and modalities. And with luck, we might ultimately arrive at theories of neural computation, which will shape not only the design of our experiments but also the very foundations of neuroscience.

SIMULATING THE BRAIN

Europe is in the process of investing over a billion euros in a project to simulate the human brain. Here, **Sean Hill**, one of the leaders in that effort, describes what the European project plans to do. **Chris Eliasmith** describes an alternative approach. The European effort starts with fine-grained details about individual neurons and the synapses between them and tries to work upward, from a detailed understanding of the logic of neural wiring toward an understanding of how that wiring underwrites behavior. **Eliasmith** starts closer to behavior, trying to build a more abstract model that captures behavior and psychological mechanisms while remaining faithful to known facts about neural organization.

WHOLE BRAIN SIMULATION

Sean Hill

Richard Feynman famously said, "What I cannot create, I do not understand." To truly understand the brain we need the tools to create it, in brain atlases, computer models, and simulations. In this essay I will talk about the Human Brain Project (HBP, www.humanbrainproject .eu), which has recently been awarded European Commission funding (~one billion euros over ten years) to provide a series of information-technology-based platforms aimed specifically at enabling a global collaboration between neuroscientists and driving innovation in neuroscience, medicine, and computing (see color plate 8). The platforms to be delivered are for Neuroinformatics, Medical Informatics, Brain Simulation, High Performance Computing, Neuromorphic Computing and Neurorobotics—each to be open for use by the global research community. These platforms are designed to bring together data about the brain, integrate it in unifying brain models, run simulations, analyze and visualize the results, and test hypotheses. The project aims to trigger a global, collaborative effort to understand the human brain, while enabling advances in neuroscience, medicine, and future computing. The primary objective is to provide the capability to build and simulate models of the entire human brain within ten years.

Creating to Understand

Virtually all neuroscientists agree that we don't yet understand the human brain. But how can we build a model of something we don't understand? Should we wait until we understand the brain to build a model of it? Or can the act of building a model itself serve as a key tool in the process of our ultimate understanding of the brain? At the Human Brain Project, we argue that the very act of organizing the data, identifying which data is missing, and evaluating how much the available data can tell us about the structure and function of the brain is essential to understanding the brain. Currently we lack the tools to search and access neuroscience data, we lack comprehensive atlases that tell us which parts of the brain have been mapped and those that have not. We also lack the tools to evaluate whether a particular piece of data is essential or irrelevant to understand the function of the brain.

Even many basic questions remain poorly understood. Why, for example, does the brain have so many different types of neurons? There are many different classes of neurons with unique electrical and morphological properties—there are likely thousands, but at this point we actually don't even know how many distinct neuron types there are. We also don't know whether it is a reasonable simplification to assume (as many computational modelers do) that the number of neuron classes ultimately comes down to just two—excitatory and inhibitory. By constructing a model that contains the *full* diversity of cell types—even for a single brain circuit—we can begin to evaluate the role of each type of neuron under different conditions and provide insight into their function in healthy brains as well as their potential impact in brain disorders and gain purchase on the essential question of which simplifications are and are not appropriate.

Another way that whole brain modeling and simulation helps create understanding is through formalizing the best theories of high-level brain function and testing their consistency with the underlying neuroscience data. Simulations are the proving ground for these high-level theories; they must be shown to be consistent with biological data in order to be valid.

There are many ways to approach modeling the brain (see also chapters by Eliasmith and Koch), but none that are yet a clear winner. Bottom-up modeling starts from data describing some of the lowest levels of detail—genes, molecules, neurons, synapses, and circuits—and aims to be faithful to all the known biological details measured in today's neuroscience laboratories. Top-down brain modeling involves developing and testing theories of brain function at a level of abstraction that typically leaves out details of individual cells and synapses. Top-down brain theories aim to provide important high-level structure scaffolding into which the biological details can be framed. Currently, high-level models of brain function can recreate some basic cognitive processes and behaviors, and bottom-up high-fidelity biophysical models can capture many details that relate gene expression products to cellular, synaptic, and even microcircuit activity. However, there is currently no model that can accurately predict the relationship between gene expression and cognition or behavior. The Human Brain Project aims to bridge the two.

Where's the Data?

A prerequisite to creating a whole brain model is the emerging new discipline called neuroinformatics—the endeavor to apply computing technology to help solve the challenges neuroscientists face in organizing, sharing, and gaining insight from their data.

Scientists have produced millions of papers and petabytes of data about the brain describing these many levels of detail—and the pace is growing even faster. Since 1990, the number of publications alone has grown from around 30,000 to nearly 100,000 per year in 2013. The number and size of large-scale datasets are also rapidly increasing—a recently produced single human brain scan consumes 1 terabyte (a thousand gigabytes) of storage—enough to fill the storage on a single laptop. The Allen Institute for Brain Science, a partner in the HBP, has led the field of neuroscience in demonstrating how to produce large-scale data—with their first atlas of gene expression of all 21,000 genes in the mouse brain. The Allen Institute produced over a petabyte (a million gigabytes) of data this year in the course of a single study to characterize mouse brain connectivity, and now, with further investment from Paul Allen, the institute is planning to acquire a tremendous amount of data about the mouse and human brain cell types, brain circuitry, and behavior over the next ten years (see chapters by Koch and Hawrylycz herein). The US Human Connectome Project is likewise producing petabytes of data describing the major connectivity pathways of the human brain and their link to individual genetics. The US BRAIN initiative will also significantly increase the amount of data available about brain structure and activity through the development of new technology and techniques. Other initiatives, like those that aim to improve the quality of scientific data through increased incentives for sharing, will also increase the sheer amount of data.

How will we make use of all this data? Already, individual scientists can only read a small fraction of the publications produced in a year, and it is a tremendous challenge for an individual scientist to understand more than a few aspects of brain structure or function. Sometimes it is difficult to know for sure even if a given experiment has already been performed or not.

A start comes from the many websites, databases, search engines, analysis services, and tools for neuroscience that have been developed and shared around the world. The decade of the brain in the 1990s brought the NIH Human Brain Project, which encouraged (and funded) neuroscientists to produce and share neuroscience data in online databases. However, the challenge of integrating this data soon became apparent—every lab used their own methods for gathering data, their own language for describing it, and their own ways of organizing it in databases with many different data formats. Thus it was extremely difficult to bring together data from different laboratories in order to gain insight about the brain. Many laboratories, for example, don't agree on the number of cell types in the brain, or even the definition of a cell type, let alone agreeing to the names of neurons!

One key ingredient, essential if not always high profile, is careful data *integration*, the precise annotation of data and management to understand how each piece of data relates to the others. Modern data integration methods, including precise metadata annotation, standard vocabularies and ontologies for describing protocols, methods and brain structures, reference coordinate spaces and "big data" analytics for performing analysis and data integration of large multiscale datasets are required. Data intensive science is a term that has come to be applied to data-driven efforts making use of modern analytic and machine learning methods to find patterns and structures in a sea of data.

The World Wide Web was invented by physicists looking for new ways to improve collaboration between scientists scattered around the planet. As this technology has been brought to bear on the lives of everyday citizens, virtually everyone has felt the impact of the information technology revolution on their lives—access to news, information, shopping; social networks; online audio and video; massive online computer services. The world has changed in the last few decades. Neuroscience needs to adopt these tools and techniques to share and integrate data about the brain.

To cope with these challenges, an international organization—the International Neuroinformatics Coordinating Facility (INCF, www .incf.org) was conceived by the Global Science Forum of the OECD and launched in 2005. This organization (in which I am now the scientific director) was given the mandate of coordinating standards and infrastructure for neuroscientists to share and integrate data globally. INCF coordinates neuroscientists around the world in agreeing on standard vocabularies for brain structures and neurons, coordinate systems for mapping the brain, mathematical modeling languages, and metadata and file format standards. INCF has worked closely with partners, including the Allen Institute, University of Oslo, Duke University, University of Edinburgh, and others, to develop a standard coordinate space for mouse brain data, dubbed "Waxholm Space" and web services to facilitate translation between mouse brain atlases. In addition, in collaboration with the Neuroscience Information Framework (NIF, San Diego) it has produced community consensus ontologies and nomenclatures for neurons and brain structures, which have been placed in a public wiki (www.neurolex.org) employing the latest semantic web technologies. INCF supports working groups of experts from around the world to produce new standards, tools, services, and guidelines for the global community. With the advent of multiple large-scale brain initiatives around the world, INCF is well positioned to help coordinate standards and infrastructure between such projects at a global scale. INCF has agreed to coordinate some of the tools for brain atlases in the HBP Neuroinformatics Platform, and the HBP will build off of INCF infrastructures and adhere to INCF standards and guidelines whenever applicable.

Next Generation Brain Atlases

One emerging tool for neuroinformatics—and key for building whole brain models—is building new types of brain atlases (see chapter by Hawrylycz). Brain atlases have been essential in making progress in understanding the organization of the brain. Traditional brain atlases were primarily anatomical, focusing on brain landmarks and features to identify brain regions. Today, new forms of brain atlases are emerging that bring together many more types of data. For example, efforts are emerging to combine data about gene expression, cellular data, and connectivity into new brain atlases. By combining multiple types of data into a single atlas, we can learn important relationships and principles; for example, how cell types from different regions connect to each other.

In the same way that early explorers relied on maps, even though they were originally coarse and error prone, neuroscientists are beginning to build atlases that anchor different types of data and help complete the layers of the complex brain maps that are gradually emerging. Ensuring data is registered and accessible through such next-generation atlases will surely be an important way of organizing and analyzing brain data. Building atlases depends on two key aspects: using jointly agreed upon vocabularies or ontologies to label data and locating the data in a standardized atlas coordinate space. For example, neuroscientists today use different words to refer to the same brain area. For example, in the visual system, *reticular nucleus of the thalamus, nucleus reticularis*, and *perigeniculate nucleus* all refer to the same brain structure—making it difficult to find all data about this single brain region. Ontologies formalize the definitions of these structures and their names (and synonyms) so that the relationships between entities are explicit. Alternatively, by annotating the data with the spatial coordinates of where it was measured, it would be associated with the volume that has been named *reticular nucleus of the thalamus*. Careful curation of data and annotating it using the next generation semantic web technologies and spatial coordinates, each piece of data will be part of a rich brain atlas integrated with a web of knowledge about the brain.

The Neuroinformatics Platform, coordinated by groups from the École Polytechnique Fédérale de Lausanne (EPFL), Karolinska Institute, University of Oslo, Forschungszentrum Jülich, Universidad Politécnica de Madrid, and Radboud Universiteit Nijmegen, will provide the tools for organizing neuroscience data in atlases that bring together collections of data about the mouse and human brains from around the world. It will provide tools for analysis of brain structure data (including electron micrographs and optical images) and functional brain data (including intracellular single cell voltage traces and parallel recordings of hundreds of neurons). It will also provide tools for *predictive neuroscience* analyses. Predictive neuroscience (described below) uses all available data and constraints to predict missing data describing the structure of the nervous system. All data, analyses, and predictions will be registered to the brain atlases. These neuroinformatics brain atlases will be the single source of curated and quality-controlled data for building brain models in the project.

Predictive Neuroscience

It is highly unlikely that the human brain will be fully mapped—with all the elements and possible interactions measured and quantified. In addition, much of the available data will actually come from other species, including mice. So a key part of the Neuroinformatics work and throughout the project is to use *predictive neuroscience*—applying theory to identify principles that can be used to fill in missing data and parameters based on the available data and knowledge. One example would be to use gene expression data for single cells combined with immunohistochemical staining to make predictions for the composition of cell types within a brain area. Another example is to identify principles that govern synapse positioning and other synaptic parameters so that a model can be constructed before all possible synaptic pathways have been completely characterized. Predictive neuroscience will be used to always provide the "best guess" for missing data, and these predictions will then be the focus for ongoing validations and further targeted experiments.

In the case of synapse positions, a recent study from the Blue Brain Project showed that it is possible to predict the distribution of synapse locations for many types of synaptic pathways in the cortex from the shapes of the axons and dendrites alone. The same study showed that there are some specific exceptions to this principle, and this finding is guiding follow-up experiments to acquire targeted data. In addition, the study revealed that the fact that every neuron has a unique individual shape is actually essential to ensure that the cortical wiring diagram remains stable and is robust to damage. The microscale connectivity is thereby used to derive principles governing macroscale cortical wiring. Thus, predictive neuroscience can serve as a key tool in filling in missing values, defining and prioritizing new experiments, and identifying key nervous system principles.

Building the Brain

The Brain Simulation Platform, led by HBP partners from EPFL, GRS-SIM, and the Royal Institute of Technology in Stockholm, will guide neuroscientists through the process of building models of proteins, cells, synapses, circuits, brain areas, and whole brains. At each step, a scientist will be prompted through the web interface to select the data, analysis methods and model-building methods necessary to construct the model. The building workflow will be populated by default parameters derived from the selected dataset, but these can be overridden so that the scientist is free to test hypotheses or examine "what-if" scenarios. For example, the workflow may populate a neural circuit with a high density of neurons taken from a normal brain, but the researcher using the platform may wish to examine the impact of reducing neuron density—as can occur during degenerative diseases such as Alzheimer's. This can be accomplished by overriding the cell density parameters for the circuit building workflow.

A catalog of all known and characterized cell types will then be used to compose and populate brain circuits. For example, in the Blue Brain Project, we have identified fifty-five morphological types of neurons in the cortical microcircuit of the rat. The morphologies of model neurons can be drawn directly from reconstructions of real neurons or synthesized by algorithms that capture the statistical properties of each type. Future work will identify relationships between gene expression and morphological properties—the goal is to predict the morphological type from transcriptomic data. This will be essential for building models of human brain circuitry where cell reconstructions are difficult to acquire and are expected to remain sparse.

In addition to the morphological types, eleven electrical types with distinct firing behaviors have been characterized in rat cortical circuitry. Each morphological type can exhibit multiple electrical behaviors—the combination of morphological and electrical types defines an additional cell type. There are 207 of these "morphoelectric" types in the cortical microcircuit. As with the morphological properties, it will be important to develop models that predict the electrical features and firing behaviors from gene expression. Learning such principles will be key to synthesizing model human neurons that have not been fully characterized experimentally.

Connectivity is determined through algorithms that compute the connectivity made possible due to physical proximity of neuronal fibers, but then it imposes additional constraints including axonal bouton density and dendritic spines. As new rules and principles governing connectivity are discovered they can be used to generate the connections.

Synaptic plasticity rules will be layered in to further refine the connectivity with specific activity patterns. Microcircuits are patterned to build brain areas, and long-range connectivity is layered in to connect them to complete whole brain circuits. At each stage of building, data is the driving force—with theory providing the guiding principles—providing constraints for optimization processes to build the functional models.

Neuroscientists using the Brain Simulation Platform will have the freedom to add their own methods of building models, drawing on the pool of data in the Neuroinformatics Platform to constrain and parameterize models in different ways in order to recreate nervous system phenomena at different scales. The goal is to provide neuroscientists with the tools to explore the impact of their own data at the systems level as well as to help define and prioritize new experiments.

The Virtuous Loop

Validation of the model is a key part of the knowledge discovery process. In the Blue Brain Project, automatic validations are run on every new model—continually evaluating the model and comparing against thousands of biological experimental findings. Attention is focused on those areas where simulation results differ from biological findings. When these two deviate, it indicates that either the available data or the model (or both) are insufficient to explain the observed biological data.

For example, in one early simulation, blocking the activity of an entire population of interneurons failed to significantly alter the network activity. Given experimental findings that clearly showed the contrary, attention was focused on the synaptic conductances assigned for this population of neurons. By seeking specific data to help more accurately model these conductances, the model was improved and the repeat of this test now showed the importance of these cells in shaping network dynamics.

Using simulated instruments, like virtual electrodes and simulated imaging techniques, the scientists can directly compare the model activity to their own experiments. In collaboration with scientists at the Allen Institute, the Blue Brain Project has developed a model of the local field potential—the signal measured when a wire electrode is placed in the brain. Simulations thus instrumented with this virtual electrode have revealed new insights into the causal link between single cell activity and the "brain-wave" phenomena measured by extracellular electrodes, including that dendritic currents may shape the local field potential.

If the model replicates experimental findings (that were not used to build the model in the first place), this is evidence the experimental findings can be explained by the measured data. However, if the model does not replicate a particular experiment, that result is also extremely informative—guiding the neuroscientist to acquire additional specific data to refine the model-building process. In either case, the model provides an important tool to test the relevance and impact of neuroscience data on a precise scientific question.

Unifying Brain Models

The core strategy within the Human Brain Project is to continually produce new releases of unifying brain models. A unifying brain model is the model that best accounts for all available data by reproducing the greatest array of experimental data. There may be branches to test new ideas and new approaches in modeling the brain circuitry, but the model that reproduces the greatest number of experimental findings while accounting for the most data will be tagged as the current release. This follows the open source model of code releases—many contributors will add improvements, but the consortium will release a single improved new version. When an improvement to a model reproduces an additional experimental finding (without diminishing previous results) it can then be accepted as a new, validated model. Any new version of a model must demonstrate that it is an improvement over previous models and reproduces more experimental results. Thus a virtuous loop is formed of continuous data integration and model refinement, resulting in a unifying model that integrates large sets of data that can be used to test new hypotheses and make specific new predictions.

These unifying models then serve as the key tools to extracting simplifying principles. Are all the details necessary to explain nervous system function? What happens if basket cells are removed, or all neuron types are collapsed to only excitatory and inhibitory simplified models? What would happen if the brain were damaged in an accident? Do we need to model the full dendritic arbor? The unifying model becomes a tool—a new integrated representation of the latest data and knowledge—with which we can test simplifying hypotheses of brain structure and function. The model becomes a tool to iteratively try different strategies of simplification and validation—always learning what the impact is on brain activity and function when leaving out a particular detail. It is these simplifications that represent the core insights and principles to be gained from this project.

Behavior: The Brain-Body-Environment Connection

Behavior is the major output of the brain. The primary output of the brain is the control of our muscles and movements, which are the basis of behavior and language. At the same time the primary input to the brain comes from our senses: vision, hearing, smell, taste, and touch. Understanding the brain will not succeed without providing sensory input, and we cannot understand the brain's ability to produce behavior without motor outputs. Thus it is essential to provide a simulated body to couple with a simulated brain. In fact the two together form a closed-loop system that will be an indispensible part of understanding the brain. The Neurorobotics Platform, led by the Technical University of Munich, will construct a platform that will provide simulated bodies and sensory organs of varying levels of detail to enable the primary goal of producing sensory input and translating motor output signals from simulated nervous system activity into virtual movements. Computational models of the retina, cochlea, and other senses will be virtually embodied in simulated models that capture head position and movement characteristics. In addition, the platform will include the simulation of virtual environments, modeling real-world physics, that can mimic traditional behavioral and cognitive testing paradigms as well as more free-ranging environments. Eventually, the models may progress to include more detailed interfaces between spinal cord and individual muscles, but they will start out in a much simpler form, for example, translating neural impulses into movement commands. These two components, simulated bodies and simulated environments, will be the core of a closed-loop simulation engine to be coupled with the large-scale brain simulations. Ultimately, simulating a virtual mouse in a virtual maze will provide an important tool for understanding the causal mechanisms of cognition, how the brain creates memories, makes decisions, and generates behavior.

Modeling Brain Disorders and Diseases

The Medical Informatics Platform, led by Centre Hospitalier Universitaire Vaudois (CHUV), EPFL, and University College London, is charged with making it possible to analyze large quantities of clinical data (across many hospitals and clinics) to build biological signatures of brain disorders and diseases. This means using machine learning to mine many thousands of patient records (without personally identifying any individual) to find characteristic disease signatures for disorders such as Alzheimer's, Parkinson's, or depression. The data will be used to find rules from the analysis of demographic data, lab tests, genetic information, and brain imaging. These rules can then be used to help diagnosis, predict prognoses, and develop new treatments for patient populations. The disease signatures that will characterize things like protein levels, brain area volumes, and cell densities will serve as tools for parameterizing whole brain models of disease. The simulations of these models can help identify potential targets for treatment.

Ethics

Simulating a whole brain raises important ethical issues. The HBP has committed substantial resources to monitoring and discussing ethical issues during the course of the project. Even if a tool that simulates the human brain is a long way off, it is important that a dialogue begin between scientists, citizens, and society at large to establish a process for guiding policy on the types of simulations that should be pursued. For example, would it be responsible to build simulations of a brain that is larger than the human brain, or how should we approach the study of conscious experience or pain? As with all scientific advances, it is important that scientists and society jointly discuss and establish responsible policies and guidelines.

Swarm Science

One goal of the Human Brain Project is to trigger and facilitate a new wave of global collaboration in neuroscience. Much as the CERN was conceived as an international resource for physics experiments, the HBP aims to serve as a global resource for neuroscience and provide a new instrument for understanding the brain.

With an Internet accessible tool to access neuroscience data, build models, and run simulations of brain circuitry, the HBP will create an opportunity for the global neuroscience community to collaborate in a way that has not previously been possible. The HBP portal will also integrate scientific social networking and make it possible to share data, analyses, models, simulations, and publications. Because each of these can be fully attributed to their contributors, it will become possible to develop new incentives for data sharing and collaboration. Impact scores can take into account which datasets, analyses, or models are the most highly used or rated. The social network graphs can be used to develop recommendations for data, models, or publications of interest to individual researchers. The portal will also support dynamic team building; it will bring together the best researchers to tackle specific challenges in understanding the brain. If successful in engaging the community, the aim is to have swarms of scientists attacking the major challenges of understanding the brain and its disorders together—in an environment where every individual will receive credit for his or her contribution.

A Global Effort to Understand the Brain

Initiatives to understand the brain and its disorders are emerging all around the globe: the Allen Institute for Brain Science, the US BRAIN Initiative, One Mind for Research, the Human Brain Project, and new initiatives forming in China, Japan, and Australia. No single initiative will accomplish the goal of understanding the brain alone; a spirit of global cooperation will be essential.

Further Reading

Hill, Sean L., Yun Wang, Imad Riachi, Felix Schürmann, and Henry Markram. 2012. "Statistical Connectivity Provides a Sufficient Foundation for Specific Functional Connectivity in Neocortical Neural Microcircuits." *Proceedings of the National Academy of Sciences of the United States of America* 109 (42): e2885–94. doi: 10.1073 /pnas.1202128109.

Human Brain Project. N.d. http://www.humanbrainproject.eu.

Markram, Henry. 2012. "The Human Brain Project." *Scientific American* 306: 50–55.

Reimann, Michael W., Costas A. Anastassiou, Rodrigo Perin, Sean L. Hill, Henry Markram, and Christof Koch. 2013. "A Biophysically Detailed Model of Neocortical Local Field Potentials Predicts the Critical Role of Active Membrane Currents." *Neuron* 79 (2): 375–90. doi: 10.1016/j.neuron.2013.05.023.

BUILDING A BEHAVING BRAIN

Chris Eliasmith

One of the grand challenges that the National Academy of Engineers identified is to reverse engineer the brain. Neuroscientists and psychologists would no doubt agree that this is, indeed, a grand challenge.

But what exactly does it mean to "reverse engineer" a brain? In general, reverse engineering is a method by which we take an already made product and systematically explore its behavior at many levels of description so as to synthesize (that is, *build*) a similar product. We attempt to identify its components and how they work, as well as how they are composed to give rise to the global behavior of the system. With systems as complex as the brain (or a competitor's silicon chip), the synthesizing step is usually carried out as a software simulation.

Reverse engineering the brain could bring many benefits. For instance, it would allow us to better understand the biological mechanisms that the brain employs and how they tend to fail in disease. At a more abstract level, reverse engineering the brain might allow us to discover effective information-processing strategies that we can import into our own engineered devices. Perhaps more surprisingly, our understanding of the basic properties of physical computation also stand to benefit from such research—neurons, after all, do not compute like a typical digital chip. In short, reverse engineering the brain will allow us to: (1) understand the healthy and unhealthy brain and develop new medical interventions, (2) develop new kinds of algorithms to improve existing machine intelligence, and (3) develop new technologies that exploit the physical principles exhibited by neural computation.

There are currently several large-scale brain simulations already being developed, each aiming to understand the actions of a million neurons or more. One project, supported by the Defense Advanced Research Projects Agency (DARPA), is IBM's SyNAPSE project, which aims to build a new kind of computer patterned after the brain. That team recently announced a five-hundred-billion-neuron simulation (the human brain has about one hundred billion neurons). The individual neurons in SyNAPSE resemble actual neurons in that they generate neural action potentials (or "spikes") to communicate, and they incorporate some elements of individual neuron physiology (although they are much simpler than their biological counterparts in that they have no spatial extent and model only a few of the many currents found in a cell). A second high profile brain model is the €1 billion Human Brain Project, which grew out of the Swiss Blue Brain project, a simulation of (thus far) one million neurons. Although the total number of neurons simulated is small by comparison to the SyNAPSE project, the Human Brain Project aspires to model individual neurons in considerable detail, capturing neuron shape, hundreds of currents, and the dynamics of neural spiking for each cell. The trade-off for this increase in biological detail is that each neuron is far more computationally costly to simulate. Compared to the few equations per neuron in the SyNAPSE project, the Human Brain Project simulations have hundreds of equations per neuron. While this level of detail can be surpassed by adding more detailed molecular dynamics or including the important contributions of glial cells, at present this degree of biological fidelity is much higher than in other large-scale models.

From a reverse engineering perspective, large-scale simulations are an important step forward. They establish the computational feasibility of simulating large numbers of components. However, existing large-scale brain simulations like SyNAPSE and the Human Brain Project lack a key ingredient for successful reverse engineering: showing how the vast array of neural components relate to *behavior*. As yet, these models do not remember, see, move, or learn, so it is difficult to evaluate them in terms of what is, arguably, the purpose of the brain.

Behavior and the Brain

My group has taken a different approach, aimed at understanding the neural underpinnings of behavior. Our most recent model, called Spaun (Semantic Pointer Architecture Unified Network), has a single eye, which takes digital images as input, and a single, physically simulated arm, which it moves to provide behavioral output (see figure 1a). Internally, its 2.5 million neurons generate neural spikes to process the input (for example, recognize and remember digits) and generate relevant output (for example, draw digits with its arm; see figure 1b). These neurons are organized to simulate about twenty out of the approximately one thousand different areas typically identified in the brain (see figure 2a). These areas were chosen to provide a suitably rich set of functions while remaining computationally tractable. The biophysical model of individual neurons that Spaun uses is quite simple. As in the SyNAPSE project, only a few equations are needed to describe each neuron. These neurons communicate using neural action potentials (spikes). When impacting a neighboring neuron's synapse, these spikes elicit a simulated version of one of four neurotransmitters (out of the tens or hundreds of different kinds) found in the brain. Again, this level of physiological and anatomical detail provides a practical compromise between computational simplicity and functionality.

One of Spaun's virtues, relative to SyNAPSE and the Human Brain Project, is its global, brain-like structure. Whereas the neurons in SyNAPSE form a largely undifferentiated, or statistically uniform mass, in Spaun they are organized to reflect the known anatomy and function of the brain. One set of neurons is modeled after those in the frontal cortices, playing important roles in working memory and the tracking of task context. Other neurons make up a simulated basal ganglia, where they help the model learn new behavioral strategies and control the flow of information throughout much of the cortex. Still others are modeled after the neurons in the occipital lobe, allowing Spaun to visually recognize handwritten digits it has never seen before. Neurons in Spaun are physiologically similar (that is, using the kinds of neurotransmitters found in that part of the brain, spiking at similar rates, and such), functionally similar (that is, are active in similar ways under similar behavioral circumstances as in the brain), and are connected in a similar manner (that is, receiving inputs from and projecting out to some of the same brain areas that a real neuron would) to neurons in the corresponding area of a biological brain (see figure 2a). For example, there are two different kinds of medium spiny neurons in the simulated basal ganglia that receive cortical projections and are inhibitory, but they have different kinds of dopamine receptors and project to different parts of the globus pallidus.

a

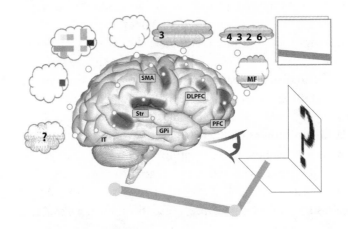

b

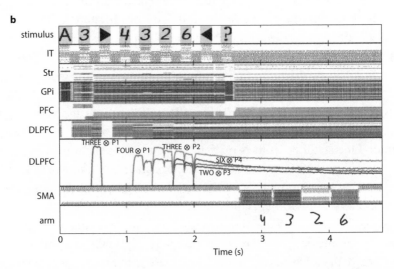

Figure 1. A serial working memory task in Spaun. a. A conceptual description of the processing Spaun performs. It is first shown a randomly chosen handwritten digit that it compresses through its visual system, allowing it to recognize the digit and map it to a conceptual representation (or "semantic pointer," SP). That representation is then further compressed by binding it to its position in the list and storing the result in working memory. Any number of digits can be shown in a row and will be processed in this manner. Once a question mark is shown, Spaun proceeds to decode its working memory representation by decompressing the items at each position and sending them to the motor system to be written out, until no digits remain. b. A screen capture from the simulation movie of this task, taken 2.5 s into the simulation time course shown in c. The input image is on the right; the output is drawn on the surface beside the arm. Spatially organized (neurons with similar tuning are near one another), low-pass-filtered model neuron activity is approximately mapped to the relevant cortical areas

To evaluate the model we compared it to a range of empirical data, drawn from both neurophysiological and behavioral studies. For instance, a common reinforcement learning task asks rats to figure out which of several actions is the best one, given some probabilistic reward (as if it were choosing between better- and worse-paying tables in a casino). Single neuron spike patterns can be recorded from the animals while they are performing this task. Spaun matches the behavioral choice patterns of the rat, but in addition, the firing patterns of neurons in the ventral striatum of both the model and the rodent exhibit similar changes during delay, approach, and reward phases of this task.

There are several examples of Spaun's neural firing patterns reproducing those found in real brains. In comparing to spiking data gathered from monkeys performing a simple working memory task, Spaun exhibits the same spectral power changes of populations of neurons (and of single neurons) while performing the same task. Similarly, by comparing to data from a monkey visual task, we have shown that the tuning of neurons in the primary visual area of the model matches those recorded in monkeys. In each case, the spiking data from the model and the animal were analyzed using exactly the same methods, to generate appropriate comparisons.

While matches to single neuron data can help build confidence in the basic mechanisms of the model, if we want to understand *human* cognition, it is often the case that such data is unavailable. As a result, in studying humans we must often rely more on behavioral comparisons. Here again, Spaun provides a good fit in many cases. For example,

---

and shown in gray scale (dark is high activity, light is low). Thought bubbles show example spike trains, and the results of decoding those spikes are in the overlaid text. For striatum (Str), the thought bubble shows decoded utilities of possible actions, and in globus pallidus internus (Gpi) the selected action is darkest. c. Time course of a single trial of the serial working memory task for four digits. The stimulus row shows input images. "A3" indicates it is performing task 3 (serial working memory), the triangles provide structure to the input, and the question mark indicates a response is expected. The arm row shows digits drawn by Spaun. Other rows are labeled by their corresponding anatomical area. Similarity plots (solid gray lines) show the dot product (i.e., similarity) between the decoded representation from the spike raster plot and concepts in Spaun's vocabulary. Raster plots in this figure are generated by randomly selecting 2,000 neurons from the relevant population and discarding any neurons with a variance of less than 10 percent over the run. Adapted from Eliasmith (2013).

a

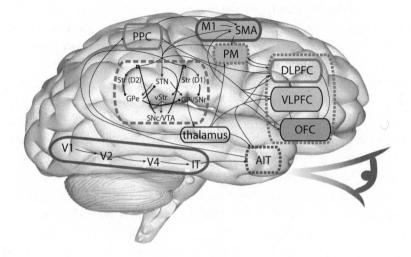

b

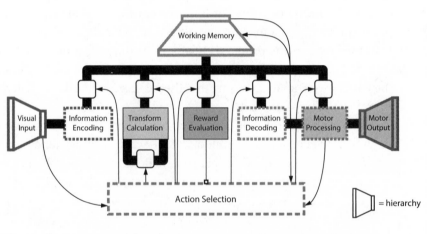

Figure 2. The architecture of the Spaun model. a. The anatomical architecture of the model (using standard anatomical abbreviations) drawn on the outline of a brain to indicate correspondences between model components and brain areas. Lines with circular endings indicate inhibitory projections. Lines with square boxes indicate modulatory connections exploited during learning. Other connections are excitatory. b. The functional organization of the model showing information flow between components. Thick lines indicate information flow between elements of model cortex, thin lines indicate information flow between the action selection mechanism (basal ganglia) and model cortex, and rounded boxes indicate elements that can be manipulated to control the flow of information within and between subsystems. The circular end of the line connecting reward evaluation and action selection indicates that this connection modulates connection weights. Line styles and fills indicate the mapping to the anatomical architecture in a. Adapted from Eliasmith (2013).

Spaun makes the same kinds and frequency of errors as humans during a serial working memory task (this task requires remembering and repeating back a list of numbers). This suggests that the neural mechanisms in the model are plausible, although evidence is indirect. Similarly, Spaun takes the same length of time per count as people do when internally counting numbers. Moreover, it also parallels people in showing an increase in the variance of the reaction of time for longer counts, reproducing Weber's famous law from psychophysics. There are many tests yet to be run, but as we continue to test the model in a variety of ways—both neurally and behaviorally—we strengthen our case that the principles we have used for reverse engineering the brain are on the right track.

This case is made significantly stronger by noting that it is the exact same model being used in each of these comparisons. Mathematical models, like Spaun, often have parameters that are tuned to match specific experimental results. This leads to the common worry that a model is "overfit" to a particular experiment or type of experiment. However, we have made significant efforts to allay this concern. For example, the decay rate of working memory is set using data from human experiments that are not included in any of the eight tasks that Spaun does. Many of the other parameters are set automatically using three principles of neural implementation that we have developed over a decade of research (Eliasmith and Anderson 2003). But, most importantly, no matter how they are set, they remain constant across all of the tasks that Spaun performs (or, more accurately, only the model can change them itself, through learning). By leaving these parameters untouched across experiments, and by testing the model against a wide variety of experiments, such concerns of overfitting become less plausible because the parameters are clearly not picked to work only under one or a few experimental conditions.

One of the central reasons for constructing such a model is to determine what it can teach us about how the brain functions. Interestingly, Spaun has generated several specific predictions that are currently being tested. For instance, the model exhibits a particular pattern of errors on question answering tasks, despite a constant reaction time in responding to questions. In particular, questions about either the identity or position of an item are more likely to be incorrectly answered the closer they are to the middle of the list. To the best of our knowledge this task has not yet been run on people, consequently it is an ideal prediction to test. Spaun has also given rise to specific neural predictions. For example, it suggests a particular pattern of similarity between the neural activity during encoding of a single item in working memory, versus encoding that same item along with other items. Specifically, the similarity of neural firing in Spaun drops off exponentially as items are added. This prediction contradicts that from other models of working memory in which the similarity stays constant. As a result, this particular prediction is an excellent test of the mechanisms and assumptions of Spaun.

In contrast to large-scale simulations that produce a lot of neural activity but little observable behavior, I would argue that Spaun is providing detailed, quantifiable insights into the organization *and function* of the brain. (Videos of many of the experiments run on Spaun can be found at http://nengo.ca/build-a-brain/spaunvideos.)

Coordinated for Flexibility

One key contribution of Spaun relative to many competing architectures is that Spaun can perform a variety of *different* behaviors, much like an actual brain. For example, Spaun can use its visual system to recognize numbers that it then organizes into a list and stores in its working memory. It can later recall this list and draw the numbers, in order, using its arm. Furthermore, Spaun can use this same visual system to parse more complex input and recognize patterns in digits it hasn't seen before. To do so, it uses the same memory system, but in a slightly different way. As well, it uses other brain areas that it didn't use in the list recall task. That is, Spaun can deploy the same brain areas in different ways depending on what task it needs to perform (see figure 2b).

This kind of "flexible coordination" is something that sets animal cognition apart from most current artificial intelligence. Animals can determine what kinds of information processing needs to be brought "online" in order to solve a given challenging problem. In other words, different, specialized brain areas are *coordinated* in a task-specific—that is, *flexible*—way to meet a challenge presented by the environment. As people, this ability comes very naturally to us, so it is an ability that we often overlook. When I switch from composing an e-mail to reading a book, to making a drink, to chasing my cat, I have coordinated many different parts of my brain in many different ways, and often with little delay in between. Because animals have evolved in a dynamic, challenging environment, this kind of behavioral flexibility is critical. In fact, Merlin Donald and others have suggested that humans are incredibly evolutionarily successful because they exhibit this kind of adaptability better than almost any other species.

One of the central goals of the Spaun project is to develop a preliminary understanding of how this kind of flexible coordination occurs in the mammalian brain. As a result, there is an important distinction in the model between midbrain and cortical regions. The midbrain regions, dominated by the basal ganglia, play a crucial role in coordinating information processing largely carried out in the cortex. So the architecture of Spaun essentially consists of an "action selector" (the basal ganglia), which monitors the current state of the cortex and determines how information needs to flow through the cortex to accomplish a given goal. However, the basal ganglia itself doesn't perform complex actions. Instead, it helps organize the cortex, so the massive computing power available there can be directed at the current problem in the right way. This allows Spaun to perform any of eight very different tasks in any order, while remaining robust to unexpected input and noise. Spaun determines what task to do by understanding its input. When it sees the letter "A" followed by a number, Spaun determines how to interpret subsequent input (for example, "A3" means that it should memorize the list of numbers it is shown next; see figure 1b).

The Benefits of Reverse Engineering

It is perhaps not surprising that in the mammalian brain, the basal ganglia have been found to be important for selecting what to do next. Problems including damage and neurodegeneration in the basal ganglia result in behaviors related to addiction, anxiety, and obsessive compulsive disorder. As well, the tremors associated with Parkinson's disease find their root in a malfunction of these areas. Consequently, understanding the mechanisms that underwrite flexible coordination have significant consequences for health.

In a similar vein, Spaun has allowed us to cast light on the cognitive decline associated with aging. There is currently a long-standing debate about whether or not the known reduction in brain cells with aging is related to the measured decline in performance on cognitive tests. The Raven's Progressive Matrices (RPM) test is a standard IQ test that has often been used to track this kind of change. The RPM test asks subjects to figure out how to complete a visual pattern of some kind. In fact, one of the tasks that Spaun performs is modeled after this test (and Spaun has been shown to perform about as well as a human of average intelligence). More recently, my lab has developed a model using the same architecture as Spaun that is able to perform the exact same test as is used on human subjects. Again, it performs about as well as average humans. Because the model has neurons, we can, for the first time, explore the causal relation between damaging those cells (as happens naturally during aging) and performance on the RPM. By running hundreds of versions of this kind of model, we can show that the performance of the models reproduces the standard "bell curve" of human populations, and that neuron loss due to aging can cause a uniform shift downward in that distribution. In short, we have been able to show how the cognitive decline due to aging could be a direct result of neuron loss.

Less obviously, understanding brain mechanisms is likely to provide us with new insights into how to build intelligent artificial systems. At the moment, most successes in machine intelligence master a single ability: machines are good at playing chess, or answering *Jeopardy!* questions, or driving a car. People, of course, can be quite good at all of these tasks. I believe this is because people can flexibly coordinate their skills in ways not currently available to machines. While most of the specific tasks that Spaun performs can be reproduced by artificial intelligence algorithms, the variety of tasks that Spaun performs is atypical of the field. Interestingly, Spaun also exhibits a nascent ability to learn new behaviors on its own (specifically, it can learn to choose different actions based on rewards in a limited manner), while preserving abilities it already has. One focus of future research on Spaun is to expand this ability to allow it to learn much more sophisticated tasks on its own, either through explicit instruction or through trial and error learning.

Building a *Physical* Brain

Even if we did understand the algorithms of the brain, it is not clear that we could usefully implement them on the computers of today. This is because the physical strategies the brain adopts for processing information lie in stark contrast to those we currently use in our computers. Silicon chips in our computing devices are engineered to eliminate uncertainty: transistors are either "on" or "off." This precision comes at the price of high power usage. Desktop computers of today typically use hundreds of watts. The brain, in contrast, uses only about 25 watts, and it performs far more sophisticated computations. And, it seems, the brain relies on highly unreliable, noisy devices: synapses fail much of the time, neurotransmitters are packaged in variable amounts, and the length of time it takes an action potential to travel down an axon can change.

Through reverse engineering, researchers have noticed these fundamental differences and have been motivated to develop "neuromorphic" silicon chips. Several of these chips arrange basic analog components of silicon chips in a manner that models of the behavior of cells in the cortex; these models have voltages with dynamics like those of neurons, and even communicate using spikes and synapses the way neurons do. Millions of such neurons can be arranged into a space smaller than a deck of cards and use less than 3 watts of power. In addition, they run in real time. This is important, since Spaun, for example, takes about 2.5 hours of real time to simulate 1 second of behavior using a digital supercomputer and kilowatts of power.

One reason these chips are promising is that many are currently fabricated with decades-old digital technology. Consequently, as they are moved to newer, already available, fabrication facilities they will be able to exploit the exponential improvement in component density. Furthermore, the limits on size affecting digital technology may not apply in the same way to neuromorphic approaches. This is because these limits are often a consequence of noise resulting from unexpected behaviors when devices get very small. Neuromorphic technology, being modeled after the noisy, stochastic brain, has faced such problems throughout its development: like the brain, neuromorphic hardware tends to be low power, analog, and asynchronous. These features tend to make the effects of noise very salient—effects usually "engineered away" in digital hardware. Consequently, the improvements in computing power and efficiency we tend to expect of digital technology may now be more readily realized by brain-based approaches.

However, one challenge in usefully employing such neuromorphic hardware has historically been a lack of methods for systematically programming noisy, low-power, highly variable hardware of this type. But, as we continue to reverse engineer neural algorithms to build large-scale brain models, we have been concurrently developing such methods. Indeed, the same techniques used to build Spaun (called the Neural Engineering Framework, or NEF; Eliasmith and Anderson 2003) have been used to program several different kinds of neuromorphic chips. Consequently, the future of both neuromorphic programming and large-scale brain modeling are intimately tied. Together I believe they will usher in a new era of low-powered, robust, flexible, and adaptive computing.

In conclusion, efforts to address the grand challenge of reverse engineering the brain are clearly underway. Large-scale models at various levels of biological detail are being developed around the world. Models like Spaun—models that connect the activity of individual neurons to behavior—are an important part of that effort, as they provide fertile, specific hypotheses that stand to significantly improve our understanding of how the brain works. While Spaun has forty thousand times fewer neurons than are in the human brain, it nevertheless provides testable and predictive ideas about neural organization and function. As such models improve—and they are likely to do so exponentially in the coming years—they will have far-reaching consequences for the development of new treatments and new technologies. These models will begin to shed light on one of the most complex physical systems we have ever encountered, and, in so doing, change our basic understanding of who we are.

References

Eliasmith, C. 2013. *How to Build a Brain: A Neural Architecture for Biological Cognition.* Oxford: Oxford University Press.

Eliasmith, C., and C. Anderson. 2003. *Neural Engineering: Computation, Representation, and Dynamics in Neurobiological Systems.* Cambridge, MA: MIT Press.

LANGUAGE

Language is uniquely human, at least in the sense of our being able to talk not just about the here and now but about the abstract, the complex, the future, and the hypothetical. Language is also uniquely difficult to study; there are no direct animal models (though birdsong can be informative), and sharp ethical limits constrain what techniques can be used. If most work in neuroscience is on vision and motor control, it's partly because those areas of the mind are more easy to study.

David Poeppel argues that the key challenge in understanding language is to bridge between a vocabulary of neural elements (such as axons and cell bodies) and a vocabulary of linguistic elements (such as nouns and verbs). He suggests that techniques such as brain imaging have not been adequate to the task but give hints into research that might take us closer. **Simon Fisher** explores an analogous set of challenges in bridging between DNA, genes, brains, and complex behavior, focusing on the case of human language. As he puts it, "we are at a watershed in genomics research, one that is set to transform multiple fields of neuroscience in unprecedented ways."

THE NEUROBIOLOGY OF LANGUAGE

David Poeppel

The Origin and Transformation of Research on the Neurobiology of Language

The ease, speed, and apparent automaticity with which we can greet a friend, follow a conversation, or read this sentence belie the considerable complexity of such seemingly effortless language tasks. Even the most elementary linguistic "event," say recognizing a single spoken word ("prose"), requires the coordination of a number of complex subprocesses (for example, analysis of the basic acoustic signal attributes, phonetic decoding, look-up/matching of the item in one's mental dictionary, retrieval of the word's pronunciation instructions, meaning, and grammatical specifications). And comprehending or producing a sentence ("Composing prose is an arduous affair") entails the subtle orchestration of dozens of underlying component operations.

Typically, one becomes aware of the internal structure and complexity of language processing only when it goes wrong; for example, following a stroke, in the case of developmental language impairment, or with severe dyslexia. In fact, it was a stroke and its medical work-up by the French neurologist Paul Broca in 1861 that played a critical role in one of the foundational insights for all of the neurosciences: the concept of *functional localization*. Broca described a language deficit and was able to relate it to a specific brain injury in the left inferior frontal lobe. This "deficit-lesion-correlation" approach has formed the basis for an impressive list of insights about brain organization for many aspects of perception and cognition—and in Broca's case, the eponymous brain region is thought to be one of the paradigmatic brain regions mediating (parts of) language.

Neuropsychological research of this type dominated the study of brain and language until the 1980s. Although the ability to discover neurobiological *mechanisms* was (and is) very restricted, this line of work has often been thoughtfully tied to psychological and linguistic research. Indeed, many of the *functional dissociations* we now take for granted originated with the careful documentation of linguistic deficits subsequent to local brain injuries. It is now beyond dispute that the parts are realized in different brain regions and by different mechanisms. However, the assignment of functions to brain regions remains a challenge. In part this has to do with questions about the granularity of the functions (what are the right level(s) of description? Language? Syntax? Noun phrase? Syntactic Constituent?), and in part it has to do with uncertainty about the biological "units" to which one ascribes a localized function (Region? Column? Microcircuit? Neuron?). A statement such as "Broca's area underpins language production" (or "speech," or "syntax," or other broad categories of linguistic experience) is not just grossly underspecified, it is ultimately both misleading and incorrect. Broca's region is not monolithic but instead is comprised of numerous subregions as specified by cytoarchitecture, immunocytochemistry, laminar properties, and so on. And domains of language such as "syntax" are similarly not monolithic but shorthand for complex suites of underlying representations and computations. It is perhaps not surprising that a brain area such as Broca's region is therefore implicated in many functions, some of which are not even particularly tied to language. For example, in addition to language-specific functions such as syntactic processing or phonology (which, one might note, are different in kind, although clearly linguistic), functional imaging studies have attributed to Broca's area the processing of hierarchically organized motor actions as well as rhythm processing. While such functions are related to language processing in a broad sense, they apply to many other domains of cognition. Future work ought to focus on "decomposing" or fractionating such complex psychological functions into putative primitive operations to account for the wide range of phenomena that are mediated by anatomically complex brain structures such as Broca's area.

Two Challenges for the New Neuroscience of Language

Current research attempts to align language research with issues at the core of systems neuroscience: detailed neurophysiological and neuro-anatomic characterizations of language processing as well as questions of neural coding. Some interesting empirical and theoretical challenges lie ahead. Two ideas for future work to grapple with are raised here, (i) a practical one and (ii) a principled one. The *practical challenge* has to do with how to conceive of the main form of data at the basis of cognitive neuroscience: maps of the brain and maps of brain activation. The *maps problem* concerns the extent to which (spatial and temporal) information about brain activity can provide a satisfactory description of the neural basis of complex brain functions. New advances in technology are making it possible to record from populations of neurons with fine spatial and temporal resolution in model systems, but insofar as the study of language is specific to humans, the field has necessarily relied on noninvasive methods. The techniques that currently dominate the field (whether spatially specialized, such as fMRI, or temporally specialized, such as EEG or MEG) predominantly characterize results in terms of spatial attributes (that is, local topographic organization, processing streams like dorsal versus ventral pathways, or networks of interconnected brain regions). Characterizing brain activity in spatial terms is intuitively straightforward and pleasing (spot A "does"/executes function X, spot B underpins function Y, and so on), often more or less correct at a broad level, and has captured the professional and the popular imaginations. While such approaches have been criticized extensively—and often with good reason, given that localization of function, even if accurate, is not equivalent to a mechanistic explanation—the fact that there exists some localization of function is an important feature of brain organization that merits explanation.

Current analyses show that some area or set of areas is selectively modulated in the context of some experimental design, and it is then argued that activation of a given region underlies, for example, "phonological processing," or "lexical access," or "syntax." Such results are, however, inevitably merely *correlational*. Even when systematic relations are implicated consistently between brain regions and certain functions, we find no explanation for why things are organized as they are, nor any sense of which properties of neuronal circuits account for the execution of function. To put it in a slogan, *localization is not explanation*.

In fact, even the highest-resolution data from (existing or to be developed) new techniques will remain inadequate—unless we succeed in decomposing language tasks into the types of primitives or computational elements that can be related to local brain structure and function. An intermediate step, in other words, is to determine theoretically well-motivated (from linguistics, psychology, computer science), computationally explicit, and biologically realistic characterizations of function to advance to better linking hypotheses. In summary, the *maps problem*—the thorough characterization of brain regions underpinning language processing—should be considered an important intermediate step, but it remains an intermediate way station that will yield correlational insights but not mechanistic explanations. The maps problem is, however, a "mere" practical limitation, that is, it is a relatively well-defined problem, and developing the linking hypotheses at the right level of granularity (discussed below) can plausibly yield satisfying descriptions of the relevant brain regions.

The *principled challenge* (as opposed to the above, practical one) deals with what we might call the "alignment" between the basic elements or primitives of language (such as syntactic units) and those of neurobiology (for example, neural circuits). What is the formal and causal relationship between the "parts list" of cognition and the "parts list" of neurobiology? The problem of mapping is the challenge of specifying the formal relations between two sets of inventories, the inventory constructed by the language sciences and that constructed by the neurosciences. The cognitive sciences, including linguistics and psychology, provide analyses of the ontological structure of various domains (let us call this the "human cognome," that is, the comprehensive list of elementary representations and operations); neurobiology similarly provides a list of the neural structures that have been identified to have functional significance. The infrastructure of linguistics—building on formally specified concepts such as *syllable* or *noun phrase* or *discourse representation*, and such—provides a structured body of concepts that allows linguists to investigate generalizations about languages that speakers bring to bear, about the course of language acquisition, about online language processing, about historical change of languages, and so on. The neurosciences—defining units of analysis such as *dendrite* or *cortical column* or *long-term potentiation*—outline the structural and functional features of the brain. But how do these ostensibly different units of analysis relate? The simplest mappings one might conceive of make little sense; there is unlikely to be any straightforward mapping between a neuron and a syllable or a cortical column and noun phrase, yet we have little idea even how to state more complex (but perhaps plausible) mappings. The fact is that we have essentially no idea how the "stuff of thought" relates to the "stuff of meat," in the case of speech and language, and much the same is true in virtually all domains of higher cognition. Bridging language and neurobiology in an explanatory fashion requires the formulation of computationally explicit linking hypotheses at the right level of abstraction.

A critical question for future research concerns at what level of abstraction to articulate such linking hypotheses. A starting point is the approach the vision scientist David Marr champions: separating computational, algorithmic, and computational levels of description. In the context of work on brain and language, the computational level of analysis is provided by linguistics and psychology, the implementational level by systems and cognitive neuroscience. A focus on the algorithmic/representational level (computer science, psycholinguistics, computational neuroscience) might provide a productive new perspective on formulating hypotheses that bridge between high-level computational and low-level implementational concerns. To provide an explicit example: many aspects of language processing, at the lexical, sentential, or discourse levels, require some (allegedly) simple operation such as concatenation (X, $Y \geq X - Y$; for example, "long paper" or "the tree" or "three blind mice"). Concatenating elements is seemingly simple and certainly ubiquitous—but has subtle properties (for example, does the conjunct, when further processed, carry the functional identity of X or Y, a property often called "headedness"?). This very straightforward operation has, as yet, no neurobiological account. It would constitute stunning progress if, in a few years, we could provide a mechanistic explanation for how neural circuits are arranged to compute "red boat" or "tasty apple." Despite the terrific progress that cognitive neuroscience of language has made in the last twenty years, mechanistic neurobiological explanations are lacking.

Some Promising Directions: Correlational Examples, with Explanatory Ambitions

Syntactic Primitives

The goals of syntactic research over the last twenty years align well with the goals of cognitive and systems neuroscience (for example, in work on computational vision, see chapter by Carandini): to identify fundamental neuronal computations that (i) underlie a large number of (linguistic) phenomena, and (ii) rely as little as possible on domain-specific properties. As a concrete example, the syntactic theory known as minimalism, developed by Chomsky and others, has formulated a two-step syntactic function called "Merge" (see above re concatenation) that separates into a domain-general computation that *combines* elements (somewhat akin to *binding*, in the context of systems neuroscience), and a probably more domain-specific computation that *labels* the output of the binding computation:

(1) Bind: Given an expression A and an expression B, bind
A,B → {A,B}
(2) Label: Given a combined {A,B}, label the complex A or
B; → {A A,B} or {B A,B}

Recent work in linguistics suggests that many of the complex properties of natural languages can be modeled as repeated applications of these Bind and Label computations. Furthermore, the formal characterization of these computations in set-theoretic terms provides a computational-level description similar to the formal characterization of neuronal computations in other domains of cognition (for example, normalization functions in vision). This new direction in syntactic theory marks a radical departure from earlier theories, which contained a large number of disparate kinds of rules and relied heavily on domain-specific properties of those rules. The time seems right for a renewed collaboration between syntacticians and cognitive/systems neuroscientists, teaming up to search for the neural circuits that subserve fundamental syntactic computations like *Concatenate/Combine/Bind* and *Label*, and ultimately the neuronal encoding of the computations themselves. Cognitive and systems neuroscientists who were dissuaded by the many rules of earlier syntactic theories may be heartened to learn that linguists have already begun to reformulate syntactic theories in terms of (putative) fundamental neuronal computations.

Speech Perception and Cortical Oscillations: Emerging Computational Principles

Recognizing spoken language requires parsing or chunking relatively continuous input into discrete units that can connect with the stored information that forms the basis for processing, or informally, words. In addition to such a parsing stage, there must also exist a decoding stage in which parsed acoustic information is transformed into representations that underpin linguistic computation. In the last decade, functional anatomic and physiological studies have identified the infrastructure of the language-ready brain. In particular, the functional anatomy of speech sound processing is comprised of a distributed cortical system encompassing regions along (at least) two streams. A ventral, temporal lobe pathway provides the substrate to map from sound input to meaning/words. A dorsal path along parietal and frontal lobes allows for the sensorimotor transformations that underlie the mapping to output representations.

Speech contains information (required for successful decoding) that is carried at multiple timescales: intonation-level information at the scale of 500–1,000 ms; syllabic information closely correlated to the acoustic envelope of speech at the scale of 150–300 ms; and rapidly changing featural (and phonemic) information at the scale ~20–80 ms. The different aspects of signals (slow and fast temporal modulations, frequency composition) must be analyzed for successful recognition. Psychophysical and neurophysiological experiments suggest that neuronal oscillations at different frequencies (delta 1–3 Hz, theta 4–8 Hz, low gamma 30–50 Hz) may provide some of the mechanisms that form the basis for parsing and decoding speech. In order to achieve the parsing/chunking of naturalistic input into manageable units, one mesoscopic-level mechanism is argued to consist of the sliding and resetting of temporal windows, implemented as phase locking of low-frequency activity to the envelope of speech and resetting of intrinsic oscillations on privileged timescales. The successful phase resetting of neuronal oscillations provides time constants (or optimal temporal integration windows) for parsing and decoding speech signals. It has been shown recently in both behavioral and physiological experiments that eliminating such oscillatory phase-resetting operations compromises speech intelligibility. Such studies connect the neural infrastructure provided by neural oscillations to well-known perceptual challenges in speech recognition. An emerging generalization suggests that acoustic signals must contain an "edge," that is, an acoustic discontinuity that the listeners use to chunk the signal at the appropriate temporal granularity.

Acoustic edges in speech are likely to play an important causal role in the successful perceptual analysis of complex auditory signals, and this type of perceptual analysis is closely linked to the existence and causal force of cortical oscillations.

Computational Neuroanatomy of Speech Production

Research on speech production is typically conducted in the context of two distinct traditions, psycholinguistics, where insights at the level of phonemes, morphemes, and phrasal level units are sought, and motor control/neural systems, concerned with kinematic forces, movement trajectories, and feedback control. These areas of research are, somewhat surprisingly, rarely linked. A standard argument regarding the disconnection states that the two approaches are focused on different levels of production tasks: psycholinguists work at an abstract and perhaps even amodal level of analysis; motor control/neuroscientists examine lower-level articulatory control processes. However, closer examination reveals provocative convergence, suggesting that both approaches have much to gain by working toward integration. For example, psycholinguistic research has documented the existence of a hierarchically organized speech production system in which planning units ranging from articulatory features to words, intonational contours, and even phrases are used. Motor control approaches, on the other hand, have emphasized the role of efference copy signals from motor commands and the role of internal forward models (the tacit "knowledge" an organism has of its action systems and effectors, providing the ability to calculate predicted outcomes of actions) in motor learning and control (see related examples in chapter by Shenoy). Integration of such notions from the two traditions has generated several hierarchical feedback control models of speech production that provide elegant links between the domains. The architecture of these models typically derives from state feedback models of motor control, but new models incorporate processing levels that have been identified in psycholinguistic research. The architecture includes a motor controller that generates forward sensory predictions. Communication between the sensory and motor systems is achieved by an auditory–motor translation system.

Building Meaning from Smaller Parts

Whether as speech, sign, text, or Braille, the essence of human language is its unbounded combinatory potential: the systems of syntax and semantics permit the composition of an infinite range of expressions from a finite (and rather limited) set of elementary building blocks. Constructing complex meanings is not just the concatenation of strings. The combinatory operations of language are subtle and invite systematic investigation. Why, for example, does every native English speaker have the clear intuition that "piling the cushions high" results in a *high pile* and not *high cushions*, whereas "hammering the ring flat" gives you a *flat ring* instead of a *flat hammer*? Questions of this type are answered by research in theoretical syntax and semantics, research areas that offer detailed cognitive models of the representations and computations that derive such complex linguistic meanings. To date, research on neurolinguistics (at least semantics) has remained almost entirely disconnected from this body of work. Consequently, our understanding of the neurobiology of the combination of words and the composition of meaning is generic and coarse. Neuroscientific work on syntax and semantics typically implicates a general network of "sentence processing regions," but the computational details have not been addressed. Recent research aims to bridge this gap: for example, some recent studies systematically vary the properties of composition to investigate the detailed computational roles and spatiotemporal dynamics of the different brain regions participating in the construction of complex meaning. The combinatory network this research implicates comprises at least the left anterior temporal lobe and the angular gyrus. Of these regions, the left anterior temporal lobe operates early (~200–300 ms) and appears specialized to the combination of predicates with other predicates to derive more complex predicates (as in *red boat*). The roles of the other regions appear to be more general and later in time (~400 ms). Contrary to hypotheses that treat natural language composition as monolithic and localized to a single region (say, Broca's area), the emerging picture suggests that composition is achieved by a network of regions that vary in their computational specificity. By connecting the neurobiology of language to formal models of linguistic representation, the work decomposes the various computations that underlie the brain's combinatory capacity. Note that such results are not yet explanatory; they correlate between putative linguistic primitives and cortical areas. However, this intermediate problem can form the basis for subsequent studies, building toward mechanistic, explanatory relations between neuronal circuitry and hypothesized elementary functions.

The study of language and its neural foundations is poised to play a pivotal role in the investigation of complex brain function. The research builds on a rich theoretical basis, established after decades of research on the computations and representations that comprise language. Moreover, the neurobiological approaches to study language are of increasingly high resolution and analytic sophistication. The significant gaps can be addressed: we are missing relevant computational analyses (at the right level of abstraction) to link language processing and neuroscience. As a consequence, the particular emphasis should be on computational linking hypotheses. Only then will the realistic goal of a new "computational neurobiology of language" be in our grasp.

Further Reading

Bemis, D. K., and L. Pylkkänen. 2011. "Simple Composition: An MEG Investigation into the Comprehension of Minimal Linguistic Phrases." *Journal of Neuroscience* 31: 2801–14.

Gallistel, C. R., and A. P. King. 2010. *Memory and the Computational Brain*. Malden, MA: Wiley Blackwell.

Giraud, A. L., and David Poeppel. 2012. "Cortical Oscillations and Speech Processing: Emerging Computational Principles and Operations." *Nature Neuroscience* 15: 511–17.

Marcus, G. 2001. *The Algebraic Mind: Integrating Connectionism and Cognitive Science*. Cambridge, MA: MIT Press.

Poeppel, D. 2012. "The Maps Problem and the Mapping Problem: Two Challenges for a Cognitive Neuroscience of Speech and Language." *Cognitive Neuropsychology* 29: 34–55.

TRANSLATING THE GENOME IN HUMAN NEUROSCIENCE

Simon E. Fisher

At the beginning of 2001, geneticists reported the initial draft sequence of the genome of *Homo sapiens,* the outcome of a huge effort that had occupied thousands of scientists across the world for more than a decade, with a price tag of several billion dollars. By 2004 this initial draft had been converted into an almost complete representation; the researchers estimated that over 99 percent of the human genome was now covered with high accuracy (less than one error in every 100,000 letters of DNA). In the years that followed, thanks to the ingenuity of a new breed of molecular biologists, we witnessed an astonishing transformation in DNA sequencing technologies. The costs, resources, and time required for reading off the letters of any person's individual genome have dropped dramatically. At the time of writing this piece (late 2013), an entire human genome, or at least a very large proportion of it, can be accurately sequenced for just a few thousand dollars in a matter of days, with cheaper and quicker "third-generation" methods about to pitch us head first into an era of personalized genomics. Clinical geneticists are already using the new sequencing tools to aid diagnosis and treatment of an array of different diseases, while many members of the general public, curious about their own hidden biology, are voluntarily sending off saliva samples to genomics companies for analysis.

In the early 1990s, when I began my research career, I was charged with decoding a specific stretch of one human chromosome; the pace of change since then is simply breathtaking. However, in some sense the rise of genomics remains a bittersweet victory. We are privileged to be the only organism on earth capable of directly reading our own genetic makeup, not only documenting that of the species as a whole, but also able to catalog the myriad variations in each individual member. At the same time, we remain woefully ignorant of what all this means for neurons, for the brain, and for human cognition and behavior. This, then, could be a key challenge laid down for future generations of neuroscientists: to properly decipher how the different words and phrases of the book that is our genome translate into a thinking, remembering, talking, feeling person. Here, I want to share a few thoughts on how the field might move toward this goal. In the interest of space, I will focus on one particular aspect of the human condition, but the issues highlighted have broad relevance. When it comes to the workings of the brain, every researcher has their favorite trait. My own passion is the enigma of human language—to my mind perhaps the most fascinating and perplexing phenomenon in all of biology. Can the novel tools of genomics help to unravel this mystery?

At first glance it seems absurd to search for explanations of human language at the level of DNA. The particular language we use cannot be encoded within our genes; it is obviously something we need to learn. An infant growing up surrounded by Japanese speakers learns Japanese, while the identical child exposed to German input during development would have become fluent in German. Without exposure to a language, a child does not acquire it. Yet at a deeper level, genes lie at the very heart of this process, helping to build a brain that is finely tuned for soaking up speech and language skills from the social environment.

This is not a new idea by any means. Over the course of many decades, multiple complementary lines of supporting evidence have been highlighted. Human children, it has been repeatedly noticed, do not need explicit instruction to acquire linguistic proficiency, and the suite of coordinated abilities they develop is astonishing in its sophistication. A few years of implicit learning, based on limited input, is enough to turn a silent infant into a master wordsmith, accumulating large vocabularies, working out how to combine them into a potential limitless number of meaningful sentences, and performing astounding feats of muscular control to convert those sentences into streams of sound, becoming adept at reverse-engineering the spoken utterances of others. Moreover, human brain mappers have made considerable progress in uncovering the neural architecture that supports these language-related functions. Although the original models early neurologists put forward are now known to be oversimplifications, the clear consensus of modern neuroscience is that there are defined sets of circuits in the human brain that play crucial roles in linguistic expression, perception, and understanding. (Whether or not such circuits are *specific* to language remains an area of debate.) Comparative approaches, investigating the cognitive and communicative abilities of other species, indicate that aspects of human language, most notably its extraordinary generative capacity, are unrivaled in the natural world. Curiously enough, chimpanzees, the animals most closely related to us, cannot remotely match the linguistic capacities of humans, despite intensive training experiments that have sought to prove otherwise.

The above observations provide indirect support for the existence of genetic influences on human speech and language capacities. What about direct evidence from the genome itself? Is it possible to go further and identify the key genes? In 2001, the same year the draft human genome was published, my colleagues and I described the first such gene, called *FOXP2*. We reported that people with damaged versions of *FOXP2* suffer from a single-gene disorder affecting speech and language. A typically developing child acquires the exquisite art of speech articulation with consummate ease, learning to make rapid coordinated sequences of movements of mouth, lips, jaw, tongue, soft palate, and larynx, and exploiting these capabilities to accurately produce novel utterances. But if a child carries a mutation disturbing *FOXP2*, these skills are disrupted. During speech, she or he makes mistakes that are inconsistent (that is, they can differ from one time to the next) and that get worse as the intended utterances become more complicated. Although intensive speech therapy can help to some degree, people with *FOXP2* mutations have persistent deficits—as adults, they still find it frustratingly hard to correctly reproduce words like "catastrophe" or "hippopotamus," doing especially badly if the words are completely novel to them (for example, pronounceable nonsense words with multiple syllables, like "perplisteronk" or "contramponist"). In-depth studies of *FOXP2* mutation cases suggest that these difficulties may stem from underlying impairments in the brain's capacity to program sequences of speech movements. Interestingly, the disorder is not confined to speech, affecting diverse aspects of spoken and written language, including production and comprehension of grammar.

Mutations that disrupt *FOXP2* are rather rare; roughly a dozen families and cases have been reported in the literature so far. Most of what is known about the associated disorder comes from intensive studies of one especially large family (tagged the "KE" family), in which fifteen relatives across three generations carry the same etiological mutation. In fact, it was the availability of such a big family that allowed us to zero in on *FOXP2* in the first place, in the days before the new genomics truly took off. But beyond the known examples of *FOXP2* mutations, there are a great many other cases of unexplained speech and language disorders in the world; indeed, in most individuals with speech and language disorders, both copies of *FOXP2* are intact. It is a safe bet, then, that other genes are acting as risk factors in a significant proportion, especially since analyses of identical and nonidentical twins show that speech and language disorders are highly heritable. The availability of cheap, quick, and easy whole genome sequencing will give geneticists unprecedented opportunities for discovering novel causal mutations and for probing more deeply into the molecular machinery that supports speech and language.

However, there is a major hitch. Over the past decade we humans have become experts in reading our own genomes, and in cataloging the genetic variations in each of us. But we are seriously lagging behind when it comes to interpreting what we are reading. To steal a crude analogy from the language sciences, it is as though a native monolingual English speaker has become adept at reading out long complex Russian texts, fluently and with perfect pronunciation, but he understands the meaning of barely a handful of the words that he is speaking.

To put it in concrete terms, consider some practical examples from the field's first forays into high-throughput sequencing of human DNA. Rather than going straight for the full three billion nucleotide letters, some researchers have stripped down the issues (and the costs) by sequencing just a subset, the ~2 percent of our genetic makeup that codes for proteins, called the exome. (Individual protein-coding parts of genes are called exons, and molecular biologists have a compulsion to coin terms carrying an "ome" suffix; this somewhat peculiar habit is even spilling over into other disciplines.) Although only a small percentage of our genome in real terms, the exome is the part that we know most intimately, the easiest to connect to biological pathways. It comprises roughly 20,000 genes that encode the amino acid sequences of proteins. These proteins display an incredibly diverse array of functions: enzymes that catalyze reactions, structural proteins that shape the cell, receptors and channels that sit in cellular membranes signaling molecules that help one cell communicate with another, regulatory factors that control the activity of other genes and proteins, and so on, all coming together to form the basic machinery of every cell. When a mutation occurs in a protein-coding gene, it may alter the amino acid sequence of the encoded protein, which could change the protein's shape. While many such mutations are benign, or have a subtle impact on properties of the protein, others can severely disturb its function, and some even yield a complete lack of the protein in question. A very large proportion of the single-gene disorders that affect the human population (cystic fibrosis, muscular dystrophy, Huntington's disease, and many others) are known to be caused by specific mutations disturbing protein-coding genes.

Beyond the exome, the remainder of the genome includes many stretches of DNA (once thought of as junk) that may have complicated roles in regulating how the exome works. While progress is being made, geneticists still struggle to make sense of this dark matter of the genome, in particular to understand the biological significance of variations that affect it. So, targeting the protein-coding genes, by sequencing a whole exome rather than dealing with an entire genome, seems a smart way of simplifying things when searching for causal variants in newly ascertained families with a disorder or trait of interest. (Although, of course, there is potential to miss the true culprit if it lies outside the exome.) When geneticists began exome sequencing in earnest, they encountered an unexpected complication. It turns out that each human individual carries a surprisingly high number of potentially deleterious mutations, typically more than one hundred. These are mutations that alter or disturb protein sequences in a way that is predicted to have a damaging effect on protein function, based on bioinformatic (computer-based) analyses. Each mutation might be extremely rare in the population, or even unique to the person or family in which it is found. How do we sift out the true causal mutations, the ones that are functionally implicated in the disorder or trait we are studying, against a broader background of irrelevant genomic change? Sometimes we can rely on a lucky convergence of findings, for example, where distinct mutations in the same gene pop up in multiple different affected families or cases. But in absence of this kind of serendipity, an alternative route is to carry out laboratory work designed to decipher the biological consequences of suspected causal mutations.

There is an impressive and expanding tool kit available for addressing this question. It is now standard practice to culture human cells at the laboratory bench, insert different variants of candidate genes, observe the consequences for the functions of the encoded proteins, and test for effects on cellular properties. Neuronal precursors can be grown in a flask, genetically manipulated, and a cocktail of growth factors can be applied to yield differentiated electrically excitable cells with the properties of functioning neurons. It has even become possible to use samples taken from non-neural tissues (for example, a skin punch or blood sample from a patient with a developmental language disorder) and turn these into distinct types of neuron-like cells in the laboratory; the procedures are not yet cheap enough for them to be routine, but this could change rapidly. Many valuable insights can be gained from cell-based analyses, the nature of which will vary depending on the class of gene being studied. At the molecular level, experiments can identify whether the gene, and the protein it encodes, is part of a network of other genes and proteins and find out how its interactions might be disturbed by candidate mutations. At the neuronal level, it is possible to use cellular systems to assess the impact of particular gene variants on key processes such as proliferation, migration, differentiation, plasticity, and programmed cell death.

There are also ample opportunities for taking such work beyond individual neurons into circuits and behavior, through the availability of animal models that can be genetically manipulated. Here the humble laboratory mouse has made a vast impact, providing a mammalian system in which it is feasible to make ever more sophisticated changes at the genetic level. Mutations can be inserted almost anywhere in the mouse genome with remarkable precision. Genes of interest can be inactivated in particular brain structures or specific neural circuits, at selected developmental time points. Their expression can be silenced at one point of life, then reactivated at another. Advances in techniques of electrophysiology and optogenetics allow researchers to directly target the functions of particular sets of neurons in a highly controlled manner in the living mouse and to make connections to behavioral outputs and cognitive performance.

Here we return to the story of *FOXP2*, since this is an area where functionally oriented experiments have made (and will continue to make) a substantial contribution. The *FOXP2* gene encodes a special type of control protein, called a transcription factor, that regulates how other genes (its targets) are switched on and off. We have taken *FOXP2* mutations observed in cases of speech and language disorder and studied their impact using human neuron-like cells that can be grown in the laboratory. For example, remember the KE family in which there are fifteen affected relatives with speech/language disorder, all carrying the same mutation in the *FOXP2* gene. This genomic variant is predicted to alter an amino acid residue at a crucial point of the encoded protein. In my lab we produced the mutant KE protein in human cells and demonstrated that it was not able to regulate target genes in the normal way. Next, by using genetic engineering to insert the KE mutation into mice, we were able to evaluate its impact on the brain at multiple levels, including initial development of neural circuitry in the embryo and subsequent functions in the postnatal nervous system. The experiments revealed early disruptive effects on the branching and process length of neurites—outgrowths that emerge from the cell bodies of neurons, eventually developing into dendrites or axons. Our mouse research also showed that the KE mutation reduces the plasticity of neural circuits, that is, their ability to modulate responsiveness to stimuli, a key aspect of learning and memory. The particular circuits that seem to be affected are ones that are already known to be important for learning to make sequences of movements—fascinating in light of the speech sequencing difficulties of humans with *FOXP2* mutations.

As our work with targeted mutations shows, even when we are interested in an aspect of behavior that appears unique to humans, we can still learn a lot through investigations of other species, especially once we have an entry point in the shape of a candidate gene. *FOXP2* appeared early in evolution—most vertebrates carry a version of this gene, leading to publications describing its neural functions in a wide range of animals, not just humans and mice (as above), but also monkeys, ferrets, rats, bats, and fish. One particularly beautiful example of how *FOXP2* has participated in the biology of other species comes from elegant neurobiological research in songbirds, which showed that the avian counterpart to *FOXP2* is crucial for the ability of a male zebra finch to learn its song. The data are consistent with the view that speech and language skills did not appear out of the blue but are built on neurogenetic mechanisms with a deep evolutionary history. This does not of course discount the idea that the neurogenetic mechanisms have been subject to modification on the human lineage, or that such modifications could have been relevant for our evolution. Indeed, there has been much interest in the demonstration of protein-coding differences between *FOXP2* in humans and chimpanzees (as well as more recent work on human-Neanderthal sequence differences elsewhere in this gene). Again, functional experiments are playing a central part in helping scientists to assess the biological relevance of the sequence changes, using the same systems (cell lines, mouse models, and such) as those used for investigating the mutations that cause disorder.

Even so, if we want to comprehensively join the dots between genes and human cognition, we cannot only depend on growing cells in the laboratory or making genetically modified animals. In recent years a new weapon has been added to the armory, one that could be powerful for making links to the human brain but that has to be wielded with care. Neuroimaging genomics involves the coupling of high-throughput DNA screening with state-of-the-art methods for noninvasive characterization of human brain structure and function, including magnetic resonance imaging, diffusion weighted imaging, and magnetoencephalography. For those laboratories with access to the appropriate equipment, it is now—in principle—straightforward to test for correlations between the specific genetic variants that people carry and aspects of their brains captured by these neuroimaging methods; examples include volumes of subcortical structures, thickness and surface area of cortical regions, strength of connectivity between different neural sites, levels of activation during a cognitive task, as well as changes in oscillatory activity. Keen neuroscientists running cutting-edge brain imaging experiments can add a genetic component to their studies simply by asking each participant to spit into a DNA collection tube. They can send these samples off for genome-wide genotyping, a technique that documents the variation at hundreds of thousands of points across all chromosomes, or they might even use them for whole genome sequencing (once the prices are low enough). Then it is a matter of testing for association between gene variants and the brain measure of interest.

Such prospects are exciting, especially because we might gain insights into neurogenetic mechanisms by studying the general population, to complement investigations of unusual cases like the KE family. For instance, those of us interested in human communicative capacities might hope to discover common gene variants that are associated with varied thickness, area, or asymmetry of key cortical regions that are linked to language. Or we could screen human cohorts for common gene variants that are correlated with altered activation of such regions during language tasks. And there are already reports in the literature of these kinds of studies. At the same time, however, some words of caution are necessary.

Already, by themselves, genomics and neuroimaging can each independently generate bewilderingly complex sets of data, with vast numbers of data points. When marrying up these two different kinds of hugely complex datasets, the risk of spurious associations becomes extremely high. It is essential to develop sophisticated approaches, both to guard against false positives and to increase chances of uncovering the real biological relationships, which are expected to have small effect sizes. Language imaging genomics is still in its infancy. The use of large sample sizes and well-constrained hypotheses—tightly focused on particular candidate genes or specific neural features—will help safeguard its future as an important tool for revealing molecular underpinnings of language.

To conclude, as I have shown in this essay, we are at a watershed in genomics research, one that is set to transform multiple fields of neuroscience in unprecedented ways. Future generations of neuroscientists are extraordinarily fortunate to have access to a wealth of data and techniques offered up by the revolution in sequencing methods. They face an exciting challenge—to distil all these As, Gs, Ts, and Cs into meaningful insights regarding the biological underpinnings of some of our most mysterious traits, such as speech and language. By taking advantage of an ever-growing tool kit for investigating gene function, it will at last be possible to bridge the mechanistic gaps between DNA, neurons, circuits, brains, and cognition.

Further Reading

Deriziotis, P., and S. E. Fisher. 2013. "Neurogenomics of Speech and Language Disorders: The Road Ahead." *Genome Biology* 14: 204.

Fisher, S. E. 2013. "Building Bridges between Genes, Brains, and Language." In *Birdsong, Speech, and Language: Exploring the Evolution of Mind and Brain*, edited by J. J. Bolhuis and M. Everaert, 425–54. Cambridge, MA: MIT Press.

Graham, S. A., and S. E. Fisher. 2013. "Decoding the Genetics of Speech and Language." *Current Opinion in Neurobiology* 23: 43–51.

Vernes, S. C., P. L. Oliver, E. Spiteri, H. E. Lockstone, R. Puliyadi, J. M. Taylor, J. Ho, C. Mombereau, A. Brewer, E. Lowy, J. Nicod, M. Groszer, D. Baban, N. Sahgal, J.-B. Cazier, J. Ragoussis, K. E. Davies, D. H. Geschwind, and S. E. Fisher. 2011. "*FOXP2* Regulates Gene Networks Implicated in Neurite Outgrowth in the Developing Brain." *PLoS Genetics* 7 (7): e1002145.

SKEPTICS

Even with the enormous investments in neuroscience that are anticipated, we may still have a lot left to figure out. Using consciousness as
an example, **Ned Block** argues that the real rate-limiting step in our
understanding may be theory, rather than data. **Matteo Carandini** cau-
tions us that it is too much to expect to be able to bridge directly from
neurophysiology to behavior, and how computation might help fill the
gap. **Leah Krubitzer** reminds us of the risks in assuming that science
can be accomplished on a timetable, and **Arthur Caplan** highlights the
practical and ethical concerns, and consequences, of a brain mapping
project, including how to fund it, what to do with the data, and how to
decide when we've succeeded. Finally, **Gary Marcus** argues that current conceptual frameworks for understanding complex cognition and
behavior are impoverished, and that in order to make progress the field
of neuroscience must significantly broaden its search for computational
principles.

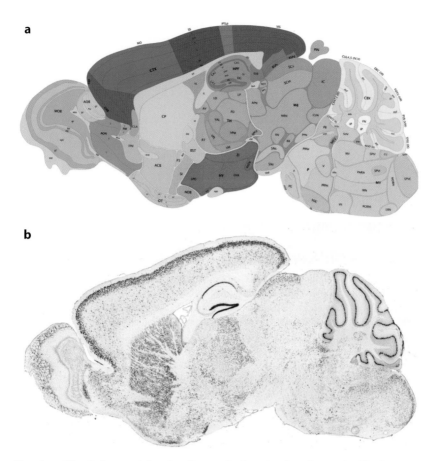

Plate 1. a. Allen Reference Atlas plate for a sagittal section (i.e., front to back) of
the mouse brain. b. *In situ* hybridization image of a calcium-binding gene (Calb 1),
showing expression in the cortex (top layer of b), striatum (left center), hippocampus
(curved shape below cortex), and cerebellum layer (top right in layer). Image courtesy
Allen Institute for Brain Science.

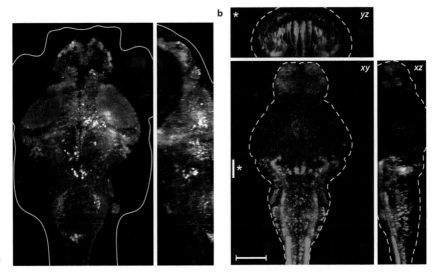

Plate 2. a. Near-simultaneous activation of many neurons across the brain of the larval zebra fish. In gray is shown the anatomy, in color code is the neural activity during one point in time. Right panel represents a maximum-intensity projection from the side, left panel from the top. b. Populations of neurons that are activated together. Two such populations, discovered by computational analysis of the raw data, are shown in green and magenta. Scale bar: 100 μm. Adapted from Ahrens, Orger, Robson, Li, and Keller (2013).

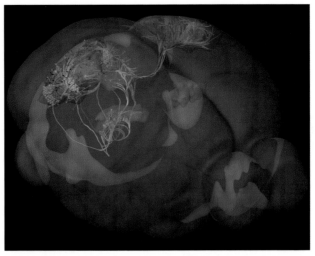

Plate 3. Connections between cortical regions involved in visual perception. Axon pathways from four distinct cortical visual areas are projected onto the Allen Mouse Brain Connectivity Atlas (http://connectivity.brain-map.org). These cortical regions are highly interconnected and are also linked with the thalamus (pink) and midbrain (purple). The front of the brain is toward the right. Courtesy of Julie Harris, Allen Institute for Brain Science.

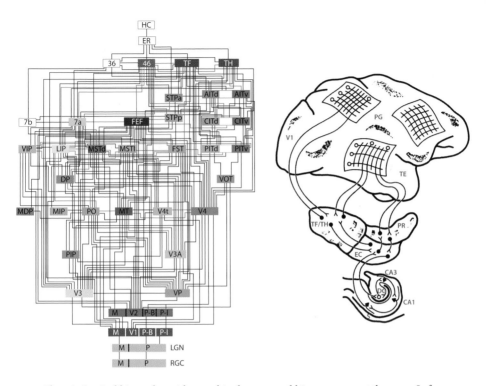

Plate 4. Cortical hierarchy, with entorhinal cortex and hippocampus at the apex. Left: Connectivity map showing hierarchy of visual areas of the cortex. Sensory cortices appear at the bottom (RGC, retinal ganglion cells; LGN, lateral geniculate nucleus). The entorhinal cortex (ER) and the hippocampus (HC) appear at the top, connected indirectly to RGC and LGN, as well as other sensory systems (not shown), all via multiple synapses. One of the goals of modern neuroscience is to understand the working principles of high-end cortices such as the entorhinal cortex and hippocampus. Adapted, by permission of Oxford University Press, from Felleman and van Essen (1991). Right: Schematic drawing of primate neocortex showing how hippocampus (bottom) and entorhinal cortex (middle) are thought to bind together representations across widespread regions of the cortex. Reproduced, with permission from AAAS, from Squire and Zola-Morgan (1991).

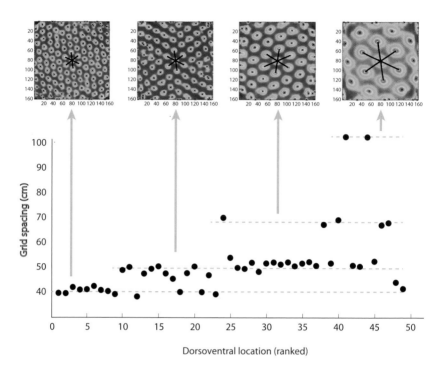

Plate 5. Modular organization of grid cells. Grid spacing is shown as a function of location of cells along the dorsal-to-ventral axis of the entorhinal cortex. Each dot corresponds to one cell. Stippled lines indicate that cells cluster into four groups with discrete values for grid spacing. Color-coded autocorrelation maps at the top show hexagonal firing patterns of individual cells representative for each of the grid modules (red is high activity; blue is low activity). Modified from Stensola et al. (2012) and Buzsaki and Moser (2013).

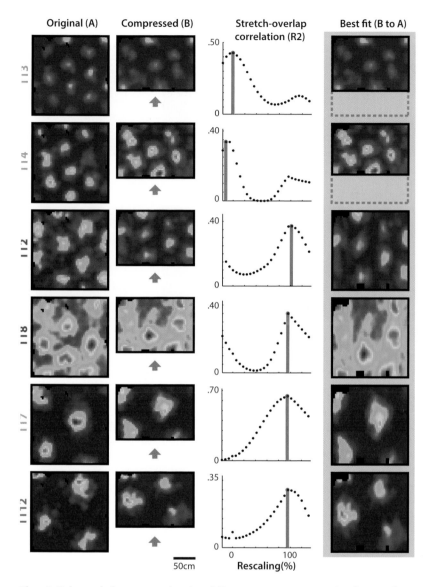

Plate 6. Color-coded rate maps showing different responses among simultaneously recorded grid cells in an experiment where the environment is changed from a square to a rectangular shape. Stretch-correlation curves to the right show spatial correlation between the compressed environment and the overlapping part of the original environment. While the two upper cells maintain their firing fields at the original location (maximal correlation at zero stretch), the fields of the four cells at the bottom are compressed in proportion to the change in the width of the environment (maximal correlation when the box is stretched back to the original shape). The upper and lower cells belong to different modules. Modified from Stensola et al. (2012).

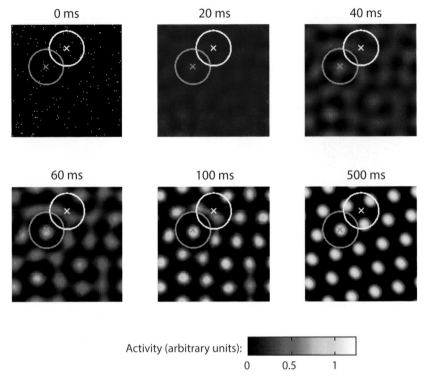

Plate 7. Computational model showing spontaneous formation of hexagonal firing patterns in grid cells. A hexagonal grid pattern forms spontaneously (here over a period of 500 ms) on a two-dimensional neuronal lattice consisting of stellate cells that have all-or-none inhibitory connections with each other. Each pixel corresponds to one cell in the network. Neurons are arranged on the lattice according to the x-y locations of their grid fields. Rings indicate area of inhibition around two example cells. Reproduced from Couey et al. (2013).

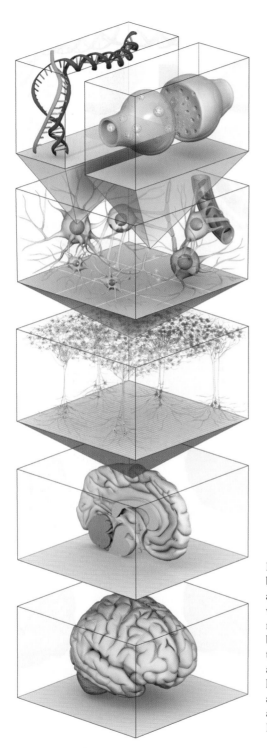

Plate 8. Bridging the levels of the brain. The Human Brain Project aims to provide neuroscientists with the tools to achieve a unified multiscale understanding of the brain—identifying the principles that link the function of the brain across levels including the molecular, cellular, circuit, region, and whole brain to cognition and behavior. Illustration © 2012 Florence Emily Cooper.

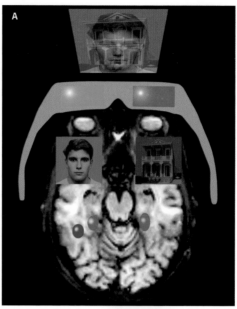

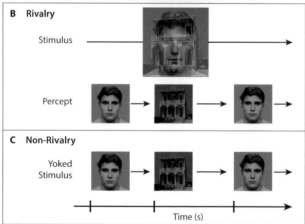

Plate 9. Panel a depicts the brain of a person looking at a superimposed red house and green face using red and green glasses, the effect of which is to transmit the image of the house to one eye and the face to the other eye. As indicated in b, the percept alternates between face and house, with only very brief mixtures, a phenomenon known as binocular rivalry. In c the subject is shown alternating pictures of a face and a house. The result is that in the crucial brain areas for perceiving faces and for perceiving places, there was no significant difference between the internally driven face/house alternation and the externally driven face/house alternation. The changing percept involved alterations in visual areas in the back of the head and also frontal areas responsible for monitoring and organizing responses. From Tong et al. (1998). With permission of Elsevier.

CONSCIOUSNESS, BIG SCIENCE, AND CONCEPTUAL CLARITY

Ned Block

With enormous investments in neuroscience looming on the horizon, including proposals to map the activity of every neuron in the brain, it is worth asking what questions such an investment might be expected to contribute to answering. What is the likelihood that high-resolution mapping will resolve fundamental questions about how the mind works? I will argue that high-resolution maps are far from sufficient, and that the utility of new technologies in neuroscience depends on developing them in tandem with the psycho-neural concepts needed to understand how the mind is implemented in the brain.

Using high school geometry, we can understand why a rigid round peg won't fit into a square hole in a board; mapping every single particle in the peg and board would be of little use without the high school geometrical account, as Hilary Putnam once noted. Similarly, a map of the activation of every neuron in the brain will be of no use without a psychological level understanding of what those activations are doing. For this reason, advocates of high-resolution mapping have advocated a "functional brain map." It is easy to add the word "functional," but massive quantities of data alone cannot produce theoretical breakthroughs in understanding the mind at a psychological level. Using the example of consciousness, I will discuss one of the obstacles to constructing a functional brain map that explains how neural activations function to underlie human psychology and how the obstacle can be circumvented without high-density brain imaging. The obstacle is the measurement problem of finding consciousness in the brain.

The Measurement Problem

The measurement problem of finding consciousness in the brain depends on the fundamental distinction between consciousness and cognition. Consciousness is what it is like to have an experience. Cognition includes thought, reasoning, memory, and decision, but all of these cognitive processes can occur unconsciously. Consciousness and cognition can causally interact, and of course cognition can be conscious, but they fall on opposite sides of a joint in nature. I will focus on the difference between conscious perception—what it is like to have a perceptual experience—and perceptual cognition—the processes in which perceptual experiences play a role in thought, reasoning, and the control of action. If an experimenter wants to know whether a subject in an experiment has consciously seen, say, a triangle, the subject has to do something, for example, say whether a triangle was present. For a subject to categorize what was seen as a triangle requires computational processes, say retrieving a representation of a triangle from memory and comparing the conscious percept with the memory trace, and there will be a further cognitive process of deciding whether to respond, and then if the decision is to respond, enumerating and deciding among candidate responses and generating a response. Further, one of the cognitive processes that can occur during a conscious percept of a triangle is a decision whether to further attend to the triangle, and subsequently the top-down attentional processes themselves. Since these cognitive processes are all in service of cognitively accessing the perceptual information and applying that information to a task, let us lump these cognitive processes all together as processes of *cognitive access*. The measurement problem, then, is how to distinguish the brain basis of consciousness from the brain basis of cognitive access.

Note that the measurement problem is distinct from David Chalmers's "hard problem" of consciousness, the problem of explaining why the brain basis of an experience of red is the brain basis of *that* type of experience rather than the experience of green or no experience at all. The hard problem depends on a prior notion of "brain basis" of the experience of red. We should be able to say *what* the brain basis of the experience of red is even if we cannot explain *why* that brain basis is the basis of that experience rather than another experience.

Why is the measurement problem a problem? Cognitive neuroscientists have identified many specialized circuits in the brain. The methodology is simple: compare the circuits that are active in, say, face perception with those that are active in other kinds of perception or when there is no perception. This methodology has resulted in the identification of the "fusiform face area" and two other linked face areas. Why can't neuroscientists just use the same idea applied to consciousness: compare what is happening in the brain during a conscious percept with what is happening in the brain during a comparable unconscious percept? One useful procedure involves presenting the subject with a series of stimuli that are at the threshold of visibility. Given the probabilistic nature of visual processing, the subject sometimes does and sometimes does not see threshold stimuli consciously. The stimuli remain the same, only the consciousness changes, so the perceptual processes common to both conscious and unconscious perception can be distinguished from the processes underlying consciousness of the stimulus. This is the "contrastive method." The problem is that, as just noted, we can only tell the difference between conscious and unconscious perception on the basis of the subject's response. So when we compare conscious with unconscious perception, we inevitably lump together the neural basis of the conscious percept with the neural basis of the response to that percept. Since the neural basis of the response underlies the very cognitive processes that I have lumped together as "cognitive access," the contrastive method inevitably conflates the neural bases of conscious perception with the neural basis of cognitive access to the perceptual content. The problem has seemed so severe that many regard it as intractable, resigning themselves to studying what I have called "access consciousness," that is, an amalgamation of the machinery of consciousness together with the machinery of cognitive access.

Further, as Lucia Melloni and her colleagues have recently shown, there are always *precursors* to a conscious state that may not be part of the neural basis of consciousness (Aru et al. 2012). For example, whether one sees a stimulus or not depends not only on fluctuations in attention but also on fluctuations in spontaneous brain activity that occur before the stimulus that may set the stage for consciousness without being part of it. To solve the measurement problem we must manage to separate consciousness from the nonconscious processes that inevitably accompany it in the situations in which we know consciousness obtains.

Indeed, the measurement problem is even thornier than I have suggested so far. Consider, for example, a type of brain injury (involving lesions in the parietal lobe) that causes a syndrome known as visuo-spatial extinction. If the patient sees a single object on either the left or the right, the patient can identify it, but if there are objects on both sides, the patient claims not to see one of the items; if the brain damage is on the right, the patient will claim to not to see the item on the left because perceptual fibers cross in feeding to the brain. However, in one such case in which a patient identified as "GK" was presented with two objects, including a face on the left that he said he did not see, Geraint Rees showed him to have activation in the relevant face area (the "fusiform face area") to almost the same degree as when he reported seeing the face. How could we find out whether GK has a conscious face experience that he does not know he has? It may seem that all we have to do is find the neural basis of face experience in unproblematic cases and ascertain whether this neural basis obtains in GK when he says he sees nothing on the left. The problem is that subjects who report seeing a face differ from those who deny seeing a face in activation of the neural basis of cognitive access to seeing a face in the frontal and parietal lobes. So it seems that in order to answer the question about GK we must first decide whether the neural basis of cognitive access to seeing a face is part of the neural basis of the conscious experience of seeing a face. And this was the question we started with.

One might wonder whether it even makes sense for GK to have a conscious face experience that he does not know about. What makes the measurement problem so problematic is the possibility that some aspect of cognitive access is actually partly constitutive of consciousness itself. If cognitive access is partly constitutive of consciousness itself, then GK could not possibly have a face experience he does not know about. If we do not solve the measurement problem, we could record every detail of activation in the face circuit and other circuits in the brain without determining whether those activations are conscious or unconscious.

The measurement problem is particularly trenchant for consciousness, but aspects of the problem arise for other mental phenomena. Masses of high-resolution data about neural activations are no use without an understanding of what the neural activations are doing at a psychological level. Once we have a theory at the psychological level, high-resolution brain data may tell us whether the theory makes correct predictions. But without the theory at the psychological level, the data are of no use no matter how high the resolution.

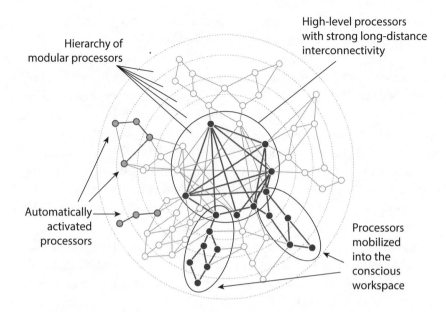

Figure 1. Diagram of the global neuronal workspace. Neural processors are symbolized by circles and connections between them by lines. Filled circles and bold lines indicate activation. The outer circles indicate sensory input, whereas the center indicates the areas in the front of the brain responsible for cognition. From Dehaene and Nacchache (2001). With permission of Elsevier.

Cognitive versus Noncognitive Theories of Consciousness

This issue—of whether cognitive access is part and parcel of consciousness—divides the field. Cognitive theories of consciousness say yes. Stanislas Dehaene, Jean-Pierre Changeux, and their colleagues (2011) have advocated a "global neuronal workspace" theory of consciousness. According to that theory, neural coalitions in the sensory areas in the back of the head compete with one another, the winners triggering "ignition" of larger networks via long-range connections to frontal areas responsible for a variety of cognitive functions. The activation of the central network feeds back to the peripheral sensory activations, maintaining their firing. Once perceptual information is part of a dominant coalition, it is available for all cognitive mechanisms and is said to be "globally broadcast" (see figure 1).

According to the global neuronal workspace theory, consciousness just is global broadcasting. Many philosophers and scientists hold versions of this view, including Sid Kouider, Daniel Dennett, and in a more attenuated form, Jesse Prinz. This is a cognitive theory of consciousness because the global workspace governs cognitive processes such as categorization, memory, reasoning, decision, and control of action. An alternative cognitive theory of consciousness David Rosenthal and Hakwan Lau (2011) hold emphasizes higher-order thought: a perception is conscious if it is accompanied by a thought about that perception. (The thought is higher order in that it is about another mental state.)

An opposed point of view, which Victor Lamme, Ilja Sligte, Annelinde Vandenbroucke, Semir Zeki, and I hold, is that activations in perceptual areas in the back of the head can be conscious without triggering global broadcasting. It is not part of our view that there can be conscious experience without any possibility of cognitive access, but only that there can be conscious experience without *actual* cognitive access. This point is shown in an experimental paradigm from Victor Lamme's laboratory illustrated in figure 2. The subject sees a circle of rectangles, then a gray screen, then another circle of rectangles. A line appears indicating the position of one of the rectangles. The line can occur with the second circle of rectangles as in A, or with the first circle as in B, or in the middle, as in C. The subject is supposed to say whether the indicated rectangle changes orientation between the first and second circle. Subjects can do this almost perfectly in B but are bad at it in A with a capacity of only four of the eight rectangles. The interesting case is C when the line appears during the gray screen. If the subjects are continuing to maintain a visual representation of all or almost all the rectangles (as they say they are doing), the difference between C and B will be small, and this is what is found. Subjects have a capacity of almost seven of the eight rectangles even when the line appears in the gray period 1.5 seconds after the first circle. The point illustrated here is that subjects can have a conscious experience of all the rectangles even though it is only possible to actually cognitively access half of them. Thus Victor Lamme and I argue that contrary to the views of those who favor a cognitive theory of consciousness, the neural basis of consciousness does not include the neural basis of actual cognitive access.

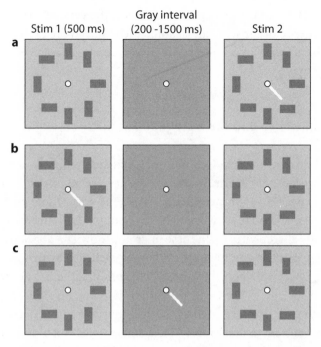

Figure 2. A perceptual task used in Victor Lamme's laboratory at the University of Amsterdam. A circle of rectangles is presented for half a second, then a gray screen for a variable period, then a new circle of rectangles. At some point in this process the subject sees a line that indicates the position of one of the rectangles. The subject's task is to say whether the rectangle at that position has changed orientation between the first and second circle of rectangles. From Lamme (2003). With permission of Elsevier.

As you might guess, this dispute has involved heavy polemics. In his 2014 book, Stanislas Dehaene says our point of view leads to dualism. He says, "The hypothetical concept of qualia, pure mental experience detached from any information-processing role, will be viewed as a peculiar idea of the prescientific era" (Dehaene 2014, 221). Of course, Lamme, Zeki, and I do not think that phenomenal consciousness has no information-processing role. We think that consciousness greases the wheels of cognitive access but can obtain without it.

The measurement problem under discussion is how it is possible for evidence to count one way or the other as between cognitive and non-cognitive theories of consciousness, given that our ability to find out whether a perception is conscious or not depends on cognitive processes by virtue of which the perception surfaces in the very behavior that provides evidence of consciousness. Some theorists have held that the measurement problem may be solved by new technology, a subject to which we now turn.

Transgenic Mice and the Optogenetic Switch

Global broadcasting involves not only feed-forward flow of activation but heavy feedback from frontal to sensory areas. Christof Koch and Nao Tsuchiya (2014) propose to use transgenic mice whose neural genes have been rendered sensitive to light, for example, by being infected with genetically altered viruses. In these mice, top-down feedback from frontal to sensory areas can be turned off optogenetically by light sources on the skull or optical fibers implanted in the brain. If there is no top-down attentional feedback there can be no "ignition" and no global broadcasting. Koch and Tsuchiya predict that without attentional feedback, the mice will be able to consciously see a single object with no distractors. On their view, top-down attention may only be required to single out an item in the visual field from other items. For example, one can detect a red "T" without top-down attention if it is the only visible object, but it takes top-down attention to detect a red "T" when the display also contains distractors: black "T"s and red "F"s.

Suppose their prediction is confirmed that the mice will be able to do a task without distractors but not when there are distractors. How we are supposed to know whether the mice whose top-down feedback has been deactivated by the optogenetic switch are doing their tasks *consciously*? Koch and Tsuchiya propose to use postdecision wagering in which the mice express their confidence in their choice by in effect betting on whether the choice is right or not. Here is how postdecision wagering works in people: the subject is given credits that are worth money. In each trial the subject makes a decision as to whether there was a stimulus present and then bets on whether that decision was right. There is a condition known as blind-sight in which destruction of parts of the lowest-level visual cortex render the subjects incapable of consciously seeing objects in the destroyed part of the visual field. Subjects can guess with very high degrees of accuracy what is presented, but they have the phenomenology of guessing, not of seeing. The blind-sight subjects bet very poorly in postdecision wagering since they have no idea which of their guesses are right, and that has suggested that betting can provide an index of conscious perception.

It turns out that animals can do something equivalent to betting to get more food pellets. And Koch and Tsuchiya say that one may be able to use postdecision wagering to test whether the optogenetic mice are consciously seeing the stimulus. High confidence would suggest conscious perception; low confidence unconscious perception. But won't the shutting off of top-down processes ruin wagering in the mice? Koch and Tsuchiya think that confidence may be mediated by different top-down processes from those involved in attention and global broadcasting and so may not be turned off by the optogenetic switch.

One way to think about this proposal is to try to imagine what it would be like to be an optogenetic mouse. Suppose you are a transgenic being whose optogenetic switch has been flipped so as to preclude top-down attention. And suppose Koch and Tsuchiya are right that you would have conscious experience. What would that experience be like? Without top-down attention, that experience would be a kaleidoscopic chaotic array of fragmentary perceptions in all sensory modalities with no sustained attention in one modality or on one thing rather than another. (Alison Gopnik has suggested that this is what it is like to be an infant in the first months of life since these infants have many more synapses and more myelination in sensory areas than in the frontal areas responsible for top-down attention.) Suppose that before the switch is flipped, you had been trained to respond to a red "T" either by itself or in a sea of black "T"s and red "F"s. Now the switch is flipped and you have a visual impression of the red "T" as part of "blooming buzzing confusion" of percepts in all sensory modalities. How much would you bet that your perception of the red "T" was accurate? It is certainly possible that the effect of the kaleidoscopic chaotic perception would be to lower one's confidence in any one percept.

Now suppose instead that the prediction of Koch and Tsuchiya is wrong—that when the optogenetic switch is flipped, it knocks out conscious perception as well as top-down attention. Without top-down signals there can be no global broadcasting. Still, the subject might be able to reliably guess whether there is a red "T" on the basis of unconscious perception as with the blind-sight patient. How would betting behavior be affected? All but one of the blind-sight patients that have been studied have had a partially blind and partially sighted field. The one human blind-sight patient whose entire visual field was blind was able to walk, with apparent confidence, through an obstacle-laden hallway. So it is hard to predict how confident a perceiver with only unconscious vision would be. In sum, betting might not correlate with consciousness once the optogenetic switch was flipped.

The upshot is that although the use of transgenic mice could make an important contribution, it would just be another line of evidence that cries out for interpretation.

Nonconceptual Representations and the Measurement Problem

Coming to grips with the measurement problem requires rethinking the basic ideas we are using. Here is a model of perception that appears in Tyler Burge's monumental *Origins of Objectivity* (2010).

Burge distinguishes between an attribute, say the circularity of the plate, and a perceptual representation, what he calls an "attributive," for example, a perceptual representation of circularity. The format of a perceptual representation is iconic and can be represented in words as "That X" where the "that" is an element that picks out an individual, the plate on the left in figure 3, and the "X" is a pure perceptual representation that picks out the circularity of the plate. The next stage to the right of the perception in figure 3 is a basic perceptual judgment in which the perceiver judges that the item is circular. Note: "That X" contains no concept, whereas "That is circular" contains the concept circular; and "That X" does not make a statement or judgment, that is, it does not say that anything is so or is the case. A basic perceptual judgment like "That is circular" is produced via the application of the concept of circularity to the percept to yield a structured propositional mental representation.

Why are we discussing percepts and concepts? Coming to grips with the measurement problem depends on understanding of the difference between two kinds of experiences: nonconceptual perceptions and conscious perceptual judgments involving concepts.

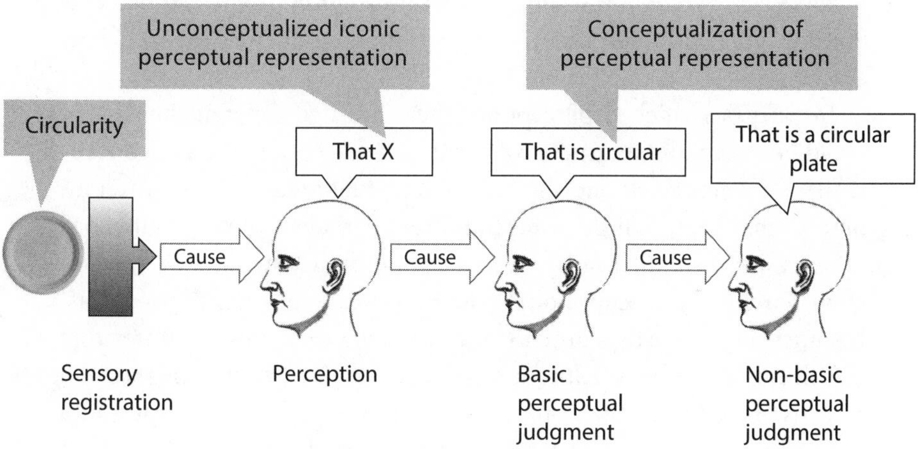

Figure 3. Burge's model of perception. © Ned Block 2013. Feed-backward influences have been omitted from this diagram. There are no top-down effects on the retina, but there are top-down influences at every other level.

What is a concept? As I am using the term "concept," a concept is a constituent of a thought or judgment that applies to something, as "circular" applies to the plate.

It is extremely important to keep separate concepts from what they are concepts of, a common confusion. For example, Bruno Latour infamously claimed that Ramses II could not have died of tuberculosis since Robert Koch discovered tuberculosis in 1882. He said, "Before Koch, the bacillus had no real existence. To say that Ramses II died of tuberculosis is as absurd as saying that he died of machine-gun fire." However, what did not exist before 1882 was not the tuberculosis bacillus, but rather the human concept of that bacillus. Many people died of tuberculosis before any humans had the concept of what killed them.

I mentioned one difference between percepts and concepts: format. Percepts are iconic; concepts are parts of thoughts or judgments that are "propositional"—they have a structure analogous to that of a sentence. Another difference is computational role: percepts are, to a first approximation, elements in a modular system, whereas concepts have a much wider role in thinking, inferring, deciding, and the like. But what is important here is not what the exact distinction is between percepts and concepts but rather that there is a joint in nature whose exact characterization is still an object of study.

In Burge's model of perception, there are two different items that could be thought of as aspects of conscious perception: the nonconceptualized percept itself and the basic perceptual judgment. A conscious percept may require little or no cognition. Perhaps a mouse could consciously perceive circularity even with no ability to think or reason about circularity. A conscious basic perceptual judgment by contrast is something that exists only in concept-using creatures, creatures that can think and reason. Although percepts can be unconscious as well as conscious, the distinction between a nonconceptual percept and a basic perceptual judgment can help in thinking about the measurement problem. One of the big advances in consciousness research in the 1990s was the realization by Francis Crick and Christof Koch that because the visual apparatus of many mammals is similar to our own, we can study perceptual consciousness in these animals even though they lack the linguistic capacities required for much of thought and reasoning. I now turn to a discussion of how the distinction may be relevant to actual experiments.

Simple Methodological Advance: Don't Ask for a Report

The familiar brain imaging pictures one sees in newspapers typically represent active brain areas. The imaging technology that produces these images—fMRI, PET, CAT—all localize spatially without much capacity to localize temporally. But in the study of conscious perception, time has proven to be as important if not more important than space. One useful technology is that of "event related potentials," or ERPs, in which electrodes placed on the scalp measure the temporally varying reaction to an event, say a visual stimulus. The brain reaction to a visual stimulus has a number of identifiable components, and researchers can and do ask which of these components correlate best with visibility of the stimulus. Stanislas Dehaene and other advocates of the global broadcasting approach have used ERP technology to find the neural basis of consciousness. And their efforts have provided evidence that the

ERP component that reflects visibility happens late in the process, when frontal concept representations have been brought into play—which is what the global broadcasting theory predicts. However, the methods Dehaene and his colleagues have used involve conceptualization of the stimulus. One study presented a target digit that was on the threshold of visibility, and the objective index of whether subjects saw it was whether the subjects could say whether the digit was larger or smaller than 5, a task that required the subject to conceptualize the seen shape in arithmetical terms and to perform an arithmetic operation, a conceptually loaded task. In another experiment, subjects had to report whether they saw the name of a number, again a task that required conceptualization of the stimulus. It is reasonable to object that what the ERP methods were revealing was not the pure percept but instead a perceptual judgment in which a concept was applied to the percept.

How can we avoid such a trap? Michael Pitts (2011) presented a series of 240 trials in which subjects saw a red ring with small discs on it. The subjects' task was to focus on the ring, looking for one of the discs to dim. Meanwhile, in the background of the ring, there were a myriad of small line segments that could be oriented randomly or, alternatively, some of the segments could be oriented so as to form one or another geometrical figure. About half the time, there was a rectangular background figure. After 240 trials of stimuli and responses about the discs were over, Pitts asked subjects to answer a series of questions that probed whether they had seen any figures in the background in the 240 trials, how confident they were about having seen these figures and what figures they saw. Those who were at least moderately confident of having seen a rectangle showed a different ERP profile from the others, and that profile differed markedly from what Dehaene and his colleagues had reported: the ERP components that correlated best with judged visibility of the rectangle came *before* global broadcasting, suggesting that subjects consciously experienced the rectangles prior to making the perceptual judgment that there was a rectangle. The activations were in perceptual areas and not in frontal areas responsible for conceptualization. The key innovation in this experiment was simple and low tech: the *relevant* conscious experience was not related to any task until *after* the perception was long gone, so the usual conflation of consciousness and cognition may not have occurred.

The idea of not asking the subject to do anything was used with an entirely different paradigm, binocular rivalry, by Wolfgang Einhäuser's lab (Frässle et al. 2014). Binocular rivalry is a phenomenon that was discovered in the sixteenth century in which two different images are presented to the two eyes. The subject's whole visual field is filled by one, then the other; the two interpretations of the world alternate with only momentary mixtures of the two images. For example, one eye may be fed a grid moving to the left and the other eye fed a grid moving to the right. The subject is aware of left motion, then right motion, then left motion, and so on. Many studies have shown that as the rivalrous percepts alternate, activations change both in the visual areas in the back of the head and in the global broadcasting areas in the front of the head, and many have taken this to support the global broadcasting theory of conscious perception. Plate 9 illustrates one of the first of these studies in which one eye is fed an image of a face and the other eye an image of a house. The percept alternates between face and house and allowed researchers to pinpoint a circuit in the brain that specializes in faces and another that specializes in houses (see color plate 9).

In the original binocular rivalry experiments, subjects reported what they were seeing by pressing a button. The Einhäuser experiment used a new method of telling when the percept shifted that did not require the subject to respond. The new method involved small eye movements that tip the experimenter off as to whether the subject is perceiving leftward or rightward motion and, in another version, changes in pupil size. The subjects' button presses validate the eye movement method, but once the method is validated the subjects do not have to do any task. The interesting result was that when there was no task there was no differential frontal brain activity. All the differences in conscious perception were in the visual and spatial areas in the back and middle of the head. The authors conclude that previous results that showed frontal global workspace changes reflected the self-monitoring required to make a response, but that when no response was required, there was little or no monitoring. Stanislas Dehaene says in his 2014 book that when "the prefrontal cortex does not gain access to . . . [a] message, it cannot be broadly shared and therefore remains unconscious" (2014, 155). But what these experiments suggest is that perceptual representations can be consciously experienced even when not actually accessed—not broadcast in the global workspace—so long as they are accessible.

This study did use new technology, but it was behavioral technology—the use of eye movements and changes in pupil size to differentiate one percept from another. These results were combined with ordinary resolution brain imaging, but ordinary resolution can be good enough when you know what you are looking for.

So we have made enormous progress in solving the measurement problem, but that progress depended on conceptual clarity, behavioral technology, and low-tech brain imaging, not expensive high-resolution brain imaging. The lesson to be drawn is that isolating consciousness in the brain may depend more on being clear about what we are looking for than on massive investments in new technology. More broadly, high-resolution data are of no use without a theory of what brain activations mean at the psychological level. When we have substantive cognitive neuroscience theories—together with the sophisticated concepts embedded in such theories—testing these theories may require Big Science. But we cannot expect the theories and concepts to somehow emerge from Big Science. To paraphrase Immanuel Kant, concepts without data are empty; data without concepts are blind; "only through their unison can knowledge arise" (Kant 1787, 75).

References

Aru, J., T. Bachmann, W. Singer, and L. Melloni. 2012. "Distilling the Neural Correlates of Consciousness." *Neuroscience and Biobehavioral Reviews* 36: 737–46.

Burge, T. 2010. *Origins of Objectivity*. Oxford: Oxford University Press.

Cohen, M., and D. Dennett. 2011. "Consciousness Cannot Be Separated from Function." *Trends in Cognitive Sciences* 15 (8): 358–64.

Dehaene, S. 2014. *Consciousness and the Brain: Deciphering How the Brain Codes Our Thoughts*. New York: Viking.

Dehaene, S., and J. Changeux. 2011. "Experimental and Theoretical Approaches to Conscious Processing." *Neuron* 70: 200–227.

Dehaene, S., and L. Nacchache. 2001. "Towards a Cognitive Neuroscience of Consciousness: Basic Evidence and a Workspace Framework." *Cognition* 79: 1–37.

Frässle, S., J. Sommer, A. Jansen, M. Naber, and W. Einhäuser. 2014. "Binocular Rivalry: Frontal Activity Relates to Introspection and Action but Not to Perception." *Journal of Neuroscience* 34 (5): 1738–47.

Kant, I. 1787. *Critique of Pure Reason*. Translated and edited by P. Guyer and A. Wood. Cambridge: Cambridge University Press, 1997.

Lamme, V. 2003. "Why Visual Attention and Awareness Are Different." *Trends in Cognitive Sciences* 7: 12–18.

Lau, H., and D. Rosenthal. 2011. "Empirical Support for Higher-Order Theories of Conscious Awareness." *Trends in Cognitive Sciences* 15 (8): 365–73.

Latour, B. 1998. "Ramses II est-il mort de la tuberculose?" *La Recherche* 307 (March): 84–85.

Pitts, M., A. Martinez, and S. A. Hillyard. 2011. "Visual Processing of Contour Patterns under Conditions of Inattentional Blindness." *Journal of Cognitive Neuroscience* 24 (2): 287–303.

Tong, F., K. Nakayama, J. T. Vaughan, and N. Kanwisher. 1998. "Binocular Rivalry and Visual Awareness in Human Extrastriate Cortex." *Neuron* 21 (4): 753–59.

Tsuchiya, N., and C. Koch. 2014. "On the Relationship between Consciousness and Attention." *Cognitive Neurosciences* 5. Forthcoming from MIT Press in a volume edited by M. Gazzaniga and G. Mangun.

Further Reading

Block, N. 2011. "Perceptual Consciousness Overflows Cognitive Access." *Trends in Cognitive Sciences* 15 (12): 567–75.

Kandel, E., H. Markram, P. Matthews, R. Yuste, and C. Koch. 2013. "Neuroscience Thinks Big (and Collaboratively)." *Nature Neuroscience* 14: 659–63.

FROM CIRCUITS TO BEHAVIOR: A BRIDGE TOO FAR?

Matteo Carandini

A fundamental mandate of neuroscience is to reveal how neural circuits lead to perception, thought, and ultimately behavior. The general public might think this goal is already achieved: when a news report says that a behavior is associated with some part of the brain, people tend to take that statement as an explanation. But neuroscientists know that most aspects of behavior result from neural circuits that are yet to established.

Clearly we need to do more work, and funders and institutions are aware of this. For instance, the University of California–San Diego has the Center for Neural Circuits and Behavior, and University College London has the Centre for Neural Circuits and Behaviour. Moreover, funding efforts such as the BRAIN initiative aim to provide critical data to achieve this result. It is right to invest in this effort, as it aims at an exciting and not unreasonable goal. But how shall we proceed? Can we indeed go directly from circuits to behavior, or might that be a bridge too far?

Imagine that instead of the brain we were trying to understand a laptop computer (figure 1a), but with the knowledge and tools available a hundred years ago. Physiologists might discover and characterize transistors, chips, buses, clocks, and hard drives. Anatomists might strive for a "connectome" of the wires across and within the chips. A furious debate, however, might divide them, as the details of wiring would differ across models (older versus newer) and across brands (AMD versus Intel). Psychologists might concentrate on general input-output properties of software applications, but those who study a business application would disagree with those studying video games. No theories, at this stage, would likely connect the hardware to the operation of the computer.

Circuits ⟷ Computations ⟷ Behavior

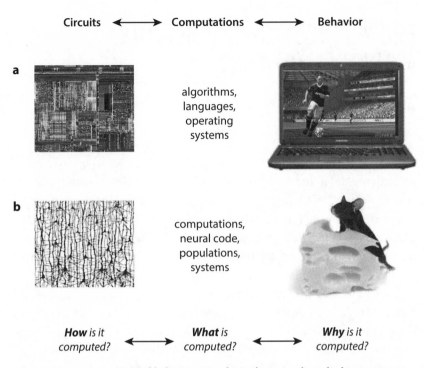

a

algorithms,
languages,
operating
systems

b

computations,
neural code,
populations,
systems

***How** is it
computed?* ⟷ ***What** is
computed?* ⟷ ***Why** is it
computed?*

Figure 1. Between circuits and behavior: David Marr's approach applied to computers and brains. a. The wiring of a fraction of an Intel microprocessor, and a laptop playing a popular videogame (FIFA 12). b. Pyramidal neurons in cortex (detail of a drawing by Ramón y Cajal), and a mouse engaged in a pleasant behavior.

What discovery would bridge this gap between circuits and behavior? It would be the realization that there is an intermediate level: the level of computer languages and operating systems. This level neatly decouples the hardware from the software. Different models and brands have different circuits but perform exactly the same computations. Different software applications are based on different combinations of instructions, but they ultimately rely on the same, finite set of computations.

Understanding these computations would allow the researchers to ask the right questions about the circuits and understand how they work. Theories about software applications, in turn, would lie on a foundation of computer algorithms without needing to speak of wires and electrical charge. In essence, grasping this intermediate level of description would explain how computers work.

In some ways, this is a tired analogy. Each generation tends to compare the brain to a complex technology of their time: a loom, a telephone exchange, a chemical plant, or a hologram. These comparisons appear comically quaint a few years later. Moreover, the brain may be more akin to a collection of special-purpose machines: the circuits for vision, olfaction, or body movement might be more tightly linked to the resulting function than in a general-purpose computer. Even so, the brain is undeniably an information-processing organ, so it may pay to compare it to our best information-processing devices.

More importantly, the computer analogy illustrates a general rule in science, which is to seek an appropriate level of description. This level is intermediate between detailed mechanism (too much reductionism) and overall function (too much holism). In physics, for instance, the equations for particle interactions become impossible to solve or even simulate once a system exceeds ten particles. So, to describe what a decent-sized piece of matter does, solid-state physicists developed remarkably successful theories operating at mesoscopic levels.

Similar examples abound in biology. For instance, we prefer to describe proteins in terms of a handful of domains rather than of thousands of amino acids. Protein domains can be identified and understood without having to refer to the precise amino acid sequence. They constitute an intermediate level that decouples the level of structure from that of overall function.

A similar approach is likely to succeed in our efforts to understand the brain: we might be able to identify an intermediate stage between circuits and behavior, a stage of computations (figure 1b). These computations are the equivalent of computer languages for brain operation and take place in the activity of individual neurons and especially of populations of neurons.

Research in recent decades, in fact, has started to reveal some of these computations. There is increasing evidence that the brain relies on a core set of standard (or "canonical") neural computations: combined and repeated across brain regions and modalities to apply similar operations to different problems.

An example of a plausible canonical neural computation in sensory systems is filtering, where neurons operate on sensory inputs by performing a weighted sum. The weights applied in this sum are called "receptive fields." This operation of filtering is performed, at least approximately, at various stages in the visual system, in the auditory system, and in the somatosensory system. It may also be at play in motor systems, where neurons can specify "force fields," assigning a weight to each body position to define a force toward a final position.

Another possible example of canonical neural computation is divisive normalization, an operation whereby neuronal responses are divided by a common factor, the summed activity of a pool of neurons. Normalization was developed to explain responses in primary visual cortex and is now thought to operate throughout the visual system and in multiple other sensory modalities and brain regions. It is thought to underlie operations as diverse as the representation of odors, the deployment of visual attention, the encoding of value, and the integration of multisensory information.

These computations are examples of bridges between circuits and behavior and of how different computations are typically considered to work in combination. For instance, a standard model of human visual detection starts with a front end made of filters, followed by a stage of divisive normalization. Filtering and divisive normalization, moreover, summarize the activity of large populations of neurons in the early visual system. As such, they have guided a multitude of experiments aimed at the underlying circuits.

Filtering and divisive normalization, however, are just two instances of plausible candidates for canonical neural computations. They are examples chosen from the author's field of expertise. Other examples, which would require other chapters and better authors to describe, include exponentiation, recurrent amplification, associative learning rules, cognitive maps, coincidence detection, top-down gain changes, population vectors, and constrained trajectories in dynamical systems. And of course one hopes that further research will identify new computations and tell us about the various ways the computations are combined in different brain regions and modalities.

Crucially, research in neural computation needs not rest on an understanding of the underlying biophysics. Some computations, such as exponentiation, are closely related to underlying biophysical mechanisms (a neuron's threshold for producing spikes). Others, however, such as divisive normalization, are less likely to map one-to-one onto a biophysical circuit. These computations depend on multiple circuits and mechanisms acting in combination, which may be different from region to region and from species to species. In this respect, they resemble a set of instructions in a computer language, which does not map uniquely onto a specific set of transistors or serve solely the needs of a specific software application.

Nonetheless, once they are discovered, neural computations can serve as a powerful guide to research in the underlying circuits and mechanisms. It is hard to understand a circuit without knowing what it is computing, be it filtering with exponentiation and divisive normalization or the detection of time differences between two sets of inputs.

Occasionally, however, it pays to proceed in the opposite direction, by starting from a circuit or biophysical mechanism and studying its computational role. This is the approach developed for instance in Christof Koch's book *Biophysics of Computation* (see also chapter by Koch, this volume). For example, studying recurrent excitation in a vertical column of cortex led to the suggestion that it may act as an amplifier and to proposals as to why amplification may be a useful computation. Similarly, discovering some sensory preferences of an inhibitory interneuron led to the suggestion that it sculpts the activity of other cells by suppressing their responses to specific stimuli.

The basic idea that one should concentrate on computation was laid out in the 1980s by David Marr in his influential book *Vision*. Marr argued that "any particular biological neuron or network should be thought of as just one implementation of a more general computational algorithm." He suggested that "the specific details of the nervous system might not matter." This might seem extreme, but it is useful as it distinguishes firmly the question of *what* is computed from the questions of *how* and *why* it is computed (figure 1).

How shall we proceed to discover and characterize more canonical neural computations, and to find out how these work in concert to produce behavior? The known neural computations were discovered by measuring the responses of single neurons and neuronal populations and relating these responses quantitatively to known factors (for example, sensory inputs, perceptual responses, cognitive states, or motor outputs). This approach clearly indicates a way forward, which is to record the spikes of many neurons concurrently in multiple brain regions, in the context of a well-defined behavior. How many neurons? Currently we record from hundreds, and new technologies will soon grow this to thousands, and perhaps soon enough, millions. Developing such technologies is precisely the goal of the BRAIN initiative, and there are already multiple hints on exciting developments on this front (see chapters by Ahrens, Koch, and Shenoy).

To guide these experiments and to interpret the resulting flood of data we will need novel theories. Ideally, these theories will establish new metaphors for the concerted activity of large neuronal populations. Great models can do that. Consider, for example, the highest success of computational neuroscience: Alan Hodgkin and Andrew Huxley's model of the action potential. This model bridged structure and function by relying not on a chemical description but on a metaphor: the equivalent electrical circuit. By extending this metaphor beyond passive membranes, it captured vast amounts of data and guided decades of research in the underlying biological hardware, i.e., voltage-sensitive ion channels (see also chapters by Freeman and Shenoy on the role of computation).

There are of course alternatives to Marr's way, and a notable one is the quest for the full diagram of the circuits of the brain, the "connectome." This diagram will undoubtedly prove useful to understand how circuits give rise to computations (the left portion of figure 1). For instance, a tiny piece of connectome was recently obtained for a piece of retina (a circuit), and it answered a long-standing question about direction selectivity (a computation). However, this approach will do little to explain how various computations are used together to produce behavior (the right portion of figure 1).

More generally, knowing a map of connections may not be as useful as one expects, especially if this map comes with no information about connection strength. For instance, we have long known the full connectome for the worm *C. elegans*, detailing the more than 7,000 connections between its 302 neurons, and yet we are hardly in the position to predict its behavior, let alone the way that this behavior is modified by learning. One of the key difficulties, highlighted by Cori Bargmann, is that a connectivity map needs to be supplemented by a moment-by-moment account of concentrations of neuromodulators, which can very quickly and radically change the function of the network. Going back to our initial analogy, those scientists who studied the laptop computer would benefit more from a manual of a programming language than from a diagram of connections between transistors in a Pentium chip (figure 1a).

Another alternative to Marr's approach is the effort to simulate brain circuits in all their glorious complexity, to obtain a "simulome" (apologies for the neologism). This approach was championed in the early 1990s with the neural simulator GENESIS and had a revival in the Blue Brain Project and most recently in the Human Brain Project (see chapter by Hill). Quoting from *The Book of GENESIS*, the key underlying hypothesis is that an "understanding of the way nervous systems compute will be very closely dependent on understanding the full details of their structure." Accordingly, one should seek "computer simulations that are very closely linked to the detailed anatomical and physiological structure" of the brain, in hopes of "generating unanticipated functional insights based on emergent properties of neuronal structure."

The problem with the simulome is that these "unanticipated insights" have not materialized. Decades since the idea was put forward, we have not discovered much by putting together highly detailed simulations of vast neural systems. Where GENESIS and other detailed neural simulators succeeded is when they concentrated on a more microscopic scale: detailed simulations of myriad items as tiny as ion channels can be necessary for understanding computation in single neurons or dendrites. However, putting all the subcellular details (most of which we don't even know) in a simulation of a vast circuit is not likely to shed light on the underlying computations. Indeed, the Blue Brain Project has hardly delivered on its initial promises, and the Human Brain Project is not poised to do any better.

In essence, while we have clear examples of success for the reductionist approach (from behavior to computations to circuits), the case still needs to be made for the constructivists' one (from circuits to computations to behavior). A similar situation is seen in other sciences: as P. W. Anderson put it, "The ability to reduce everything to simple fundamental laws does not imply the ability to start from these laws and reconstruct the universe."

Luckily, there is a strong sense that the levels of the subcellular and of the network are neatly decoupled. For instance, work by Eve Marder and others has shown that very similar patterns of cellular and network responses (and therefore very similar computations) can be obtained with wide differences in biophysical details. Conversely, work by Zachary Mainen and Terrence Sejnowski has shown that small changes in biophysical details can lead to wide differences in cellular properties (and therefore in computations). This decoupling of levels gives us hope that we will indeed understand the relationships between circuits and behavior. If, instead, understanding behavior requires understanding a myriad of relationships between molecules, channels, receptors, synapses, dendrites, neurons, and so forth, we have little hope of success.

To conclude, the gap between circuits and behavior is too wide to be bridged without an intermediate stage. Following on the basis laid by Marr, it seems evident that this stage is one of computation. Neuroscientists have already identified some computations that appear to be canonical: repeated and combined in different ways across the brain. Hopefully new experiments, new technologies, and new theories will soon identify an even wider array of computations and give us more concrete examples of how these are combined to determine behavior. Of course, this view does not advocate separating those who study circuits from those who study behavior. Rather, it argues that researchers of circuits and of behavior go furthest when they speak a common language of neural computation, a language that we are only beginning to learn.

Note

An earlier version of this chapter was published in March 2012 in *Nature Neuroscience* 15 (4): doi: 10.1038/nn.3043, by Nature Publishing Group, a division of Macmillan Publishers Limited.

Further Reading

Anderson, P. W. 1972. "More Is Different." *Science* 177: 393–96. doi: 10.1126/science.177.4047.393.

Bargmann, C. I. 2012. "Beyond the Connectome: How Neuromodulators Shape Neural Circuits." *Bioessays* 34: 458–65.

Bower, J. M. 1998. Chapter 11 in *The Book of GENESIS: Exploring Realistic Neural Models with the GEneral NEural SImulation System*, edited by J. M. Bower and D. Beeman. Santa Clara, CA: Telos.

Carandini, M., and D. J. Heeger. 2011. "Normalization as a Canonical Neural Computation." *Nature Reviews Neuroscience* 13: 51–62. doi: 10.1038/nrn3136.

Harris, K. D., et al. 2010. "How Do Neurons Work Together? Lessons from Auditory Cortex." *Hearing Research*. doi: 10.1016/j.heares.2010.06.006.

Koch, C. 1999. *Biophysics of Computation*. Oxford: Oxford University Press.

Laughlin, R. B., and D. Pines. 2000. "The Theory of Everything." *Proceedings of the National Academy of Sciences USA* 97: 28–31.

Laughlin, R. B., D. Pines, J. Schmalian, B. P. Stojkovic, and P. Wolynes. 2000. "The Middle Way." *Proceedings of the National Academy of Sciences USA* 97: 32–37.

Lazebnik, Y. 2002. "Can a Biologist Fix a Radio?—Or, What I Learned While Studying Apoptosis." *Cancer Cell* 2: 179–82.

Marr, D. *Vision*. 1982. San Francisco: W. H. Freeman.

Prinz, A. A., D. Bucher, and E. Marder. 2004. "Similar Network Activity from Disparate Circuit Parameters." *Nature Neuroscience* 7: 1345–52. doi: 10.1038/nn1352.

LESSONS FROM EVOLUTION

Leah Krubitzer

When invited to contribute to this book, *The Future of the Brain: Essays by the World's Leading Neuroscientists*, I agreed for two reasons. The first and most obvious is that I study the brain. However, as an evolutionary neurobiologist I am more interested in its past than in its future. The second reason is based on pure vanity; who could resist agreeing to be included among the "world's leading neuroscientists"? In this essay I reflect on a few important things I've come to appreciate about brain function and evolution, where I think we should direct our future energies in trying to understand the brain, and end with a brief assessment of our current ability to predict future brain evolution.

One of the first and most important lessons I have learned as a neuroscientist is that in order to understand how complex brains evolve and work, it is not enough to study only complexly organized brains. As a young graduate student, I was interested in why humans behave the way that they do, how the brain generates this behavior, and how both the brain and behavior evolve. Although much of my graduate work was on the brains of nonhuman primates, I ultimately concluded that to truly understand how complex brains evolved, looking at our close relatives like monkeys would never be enough. Although monkey brains are extremely complex, there are important insights to be gleaned from a wide variety of species. For example, we know from comparative studies in mammals that the neocortex, the part of the brain involved in perception, cognition, and volitional motor control, varies dramatically in size and the number of interconnected cortical fields (functional subdivisions of the neocortex) in different species. Comparative studies indicate that a large neocortex with multiple parts evolved in primates, including humans, but also evolved independently in other lineages such as cetaceans (whales and dolphin). In order to appreciate how these types of complex brains evolved, I felt it was critical to appreciate how the neocortex of early mammals was organized and then determine the types of alterations that were made to the brains of their descendants. Thus I ventured to Australia where I could study mammals whose ancestors branched off early in evolution (monotremes and marsupials) in the hope that they would have retained some primitive features of neocortical organization inherited from our early ancestors over two hundred million years ago. While in Australia I found that monotremes and marsupials have the same basic plan of neocortical organization that all species possess, and that this plan has been elaborated in different lineages. Thus every living mammal, including humans, has aspects of neocortical organization and connectivity that were inherited over two hundred million years ago from the common ancestor of all mammals.

The second important lesson I learned is that unusual mammals can tell us a lot about the rules of brain construction and brain/body relationships. Comparative studies on animals that possess extreme specializations like the duck-billed platypus, star-nosed mole, or echolocating bat provide important insights about the human brain. For example, the duck-billed platypus has a highly specialized bill with electrosensory receptors and uses this specialized body part for navigating, mating, and prey capture in the water. This specialized body part is associated with a number of brain features, such as cortical magnification or the amount of cortex devoted to processing inputs from a specific body part. The platypus is unique in the extraordinary magnification of its bill; about 90 percent of its somatosensory cortex is devoted to the bill representation. These body specializations in mammals are also associated with the types of stimuli that neurons respond to and alterations in the connections of the brain. Studies on animals that are highly specialized also inform us about the importance of use of this specialized body morphology in constructing the brain during development and the dynamics in shaping the neocortex as an organism matures to adulthood. If we consider human specializations in this same light, we would conclude that specializations of the vocal tract and oral structures associated with speech production have a large portion of the neocortex devoted to processing these inputs, and these areas have altered connections associated with these specializations—and, they do. As Ted Bullock elegantly articulated in his *Science* essay, "Comparative Neuroscience Holds Promise for Quiet Revolutions," comparative studies are important in revealing the *roots* or evolutionary history of brain organization, the *rules* of construction of brains and the constraints under which the nervous system develops and evolves, and the *relevance* or general principles of brain organization. Thus while we may be interested in how complex brains like those of humans arose, we must admit that most insights about general rules of construction and general principles of neocortical function come from the brains of other mammals.

The third important lesson is that the brain does not develop or function in a vacuum. For years I used comparative analysis in a variety of mammals to determine how the brain, particularly the neocortex, was modified throughout the course of evolution, and the factors that contribute to aspects of the cortical phenotype such as organization and connectivity. I was extremely "braincentric" when considering these issues, and this was due, in part, to my early training. Although I worked on multiple species as a graduate student, my experiments were restricted to listening to and looking at the brain using electrophysiological recording techniques and neuroanatomical techniques, respectively. My point is that I never seriously considered other parts of an animal except its brain. Perhaps one of the biggest revelations in my career came when I began a postdoc in Australia and had to *catch* the animals I worked on—a moment I still remember with clarity: late at night rowing a boat in murky waters, hoisting gill nets and hoping like hell there would be a platypus caught in the net. I vividly recall marveling over the texture and composition of its bill, its tiny eyes, its webbed paws and unbelievably thick, water-resistant fur, and wondering what it would be like to be a platypus. When I discovered the extraordinary amount the neocortex devoted to processing inputs from the bill, I finally realized my curiosity never could be satisfied. Although my brain shares a number of features of organization with the platypus, I don't have a hydrodynamically constructed body like a platypus, nor massive inputs from mechanosensory and electrosensory receptors on a bill pouring into my brain. Brains do not operate in isolation but are embedded in a body, often containing specialized sensory receptor arrays, and the whole animal develops and evolves in a context of both animate and inanimate objects, conspecifics (same species), and heterospecifics (other species), all of which are constrained by the laws that govern matter and energy on our planet.

The fourth important lesson learned was that genes are not everything. It is becoming more and more apparent that epigenetic mechanisms—which alter transcription or expression of genes—are critical for constructing a brain that is highly adapted to the context in which it develops and in which the animal will ultimately live.

Conrad Waddington first used the term *epigenetics* in the middle of the last century in an effort to explain cellular differentiation during development. If there is a one-to-one correspondence between DNA and the phenotype, then every somatic cell in the body (which contains exactly the same genotype) would be identical. Instead, the phenotypes of cells vary from brain cells (neurons) to liver cells. Because of this, Waddington proposed that the mechanisms through which a genotype produces a phenotype should be termed *epigenetics.*

Considering that cellular phenotypes undergo dramatic plasticity during development while the genotype of these cells remains stable implicit in Waddington's definition is the notion that a phenotype can be altered without changes to the genotype. Thus during the course of development, epigenetic mechanisms (such as DNA methylation, a biochemical process that reduces gene expression in specific portions of the brain and body) allow cells with the same DNA to differentiate and divide, passing on those alterations in gene function, not explained by alterations in DNA sequence, to daughter cells. If we expand this concept to take into account the fact that an organism does not remain static throughout the lifespan, but rather it dynamically responds to social and environmental contexts, then epigenetic mechanisms might also mediate the adaptability of brain and behavior to the environment. Recent work from the laboratories of Michael Meaney and Frances Champagne indicates that variation in early development induces epigenetic variation (in DNA methylation for example) and may serve as a mechanism for developmental plasticity. For example, alterations in nutrition, stress, and maternal care early in life can trigger these epigenetic mechanisms and generate anatomical and functional changes to the brain and body, which alters behavior of the offspring. These alterations in behavior can be sustained across generations via epigenetic effects on portions of the neuroendocrine system, or in some instances persist through epigenetic effects on the germ line.

The dramatic role that epigenetic mechanisms play in shaping brain and behavior is well exemplified in humans. Anatomical alterations to the hand necessary for complex bimanual dexterity; to the supralaryngeal tract necessary for speech production; and to the inner ear, which amplifies frequencies associated with human speech, were present well before these behaviors that we attribute to modern humans were expressed within the population. Thus the anatomical underpinnings for complex human behaviors were present in our very early ancestors and those of our Neanderthal cousins, but complex behaviors like language and sophisticated and precise tool use (generated by the neocortex) were shaped by the social and cultural context in which individuals developed, rather than traditional evolutionary mechanisms. We know from our own work and from that in other laboratories that context, which can be considered as complex and dynamic patterns of incoming sensory information available to developing brains, can alter neocortical connectivity, functional organization, and the resultant behavior of an individual. Remarkably, it is possible to dramatically alter "normal" brain connectivity and function by altering the patterns of stimuli experienced during development and over a lifetime.

This leads to my fifth revelation: there is no single or optimal way to build some feature of brain organization. For years I searched for "the way" in which some aspect of the cortical phenotype could be altered during the course of evolution. For example, what is the way in which the size of cortical fields is altered? What is the way in which cortical connections change? What is the way in which cortical fields are added? Studies of molecular development that examine genes intrinsic to the developing neocortex have demonstrated how these genes (and genetic cascades) can alter cortical field size, location, and connectivity. Interestingly, these same features of organization can be altered by the sensory driven activity that the developing organism is exposed to. Because cortical field size and connectivity can be changed through different mechanisms, this implies that in a given lineage, some aspect of brain organization owes its particular phenotype to genes, activity-dependent mechanisms, or some combination of both. However, a similar phenotype in a different mammal may have arisen by a very different combination of these factors.

From a personal rather than scientific standpoint, the final important thing I've learned is don't be taken in by the boondoggle, don't get caught up in technology, and be very suspicious of "initiatives." Science should be driven by questions that are generated by inquiry and in-depth analysis rather than top-down initiatives that dictate scientific directions. I have also learned to be suspicious of labels declaring this the "decade of" anything: The brain, The mind, Consciousness. There should be no time limit on discovery. Does anyone really believe we will solve these complex, nonlinear phenomena in ten years or even one hundred? Tightly bound temporal mandates can undermine the important, incremental, and seemingly small discoveries scientists make every day doing critical, basic, nonmandated research. These basic scientific discoveries have always been the foundation for clinical translation. By all means funding big questions and developing innovative techniques is worthwhile, but scientists and the science should dictate the process. There are numerous examples where individuals, rather than top-down initiatives, worked to progressively cure or prevent diseases or uncover important and fundamental principles of biology. Some of these include Jonas Salk's vaccine for poliomyelitis; Santiago Ramón y Cajal's discoveries on the anatomical structures of neurons and his articulation of the neuron doctrine; and of course Charles Darwin's detailed observations that led to the theory of evolution through natural selection, which is now the cornerstone of all of biology.

Of course most of these lessons learned during my career have been well documented by erudite neuroscientists well before me. However, this personal synthesis has shaped my own science and the evolution of my thoughts, and it certainly plays a heavy hand in where I believe we should direct our future energies as neuroscientists. First, I think that revealing the relationships between multiple levels of organization, from genes to neurons to cortical maps to behavior, is critical. This will require those of us working in science to step out of our individual scientific comfort zones and to consider levels of organization larger and smaller than the one at which we personally work. Our quest for understanding species differences must move well beyond comparative genomics and approaches that seek simple genetic explanations for complex phenomena such as language, autism, or schizophrenia. In our enthusiasm for genetics, we often seem to have sidestepped systems neuroscience, cognitive neuroscience, social science, and whole animal physiology, prematurely narrowing our search to uncover unrealistically direct gene-to-complex-behavior relationships. As noted above, context is extremely important, and in terms of human brain organization and function, culture appears to have played a pivotal role in shaping the human brain and modern human behavior.

Given the enormous role of social and cultural context in human brain organization and function, to predict the future evolution of the brain—where our own brains might be a hundred or thousand or a million years from now—would require us to predict the direction of social, economic, and technological changes to our current culture. We also need to consider the physical changes in the environment like global temperature, the types of food we eat, the chemical treatment of our water, alterations in our form of locomotion, and our movement away from traditional tool use to automation and skills that require more unique movements of our digits, all of which may shape our future body morphology, physiology and metabolism. In short, you can't predict future brain organization in isolation, but must consider the multilayered context in which the brain develops.

Having said this, I contend that understanding the history of brain evolution does provide powerful insight into understanding the types of alterations that can be made to brains in the future. Evolution of the neocortex can be considered, to some extent, as an ever-diminishing set of options. Genetic contingencies and pleiotropy (a single gene has multiple, seemingly unrelated effects) place formidable constraints on brain development as do the laws of physics, and comparative studies demonstrate that the types of changes that have been made to the neocortex through the course of evolution are limited. While no one can predict the exact phenotype that the next million years of human evolution will produce, one can infer the types of alterations that can be made to the human brain, as well as alterations that are improbable. One can also predict with a high degree of confidence that concrete anatomical and physiological alterations that generate complex behavior will be due to alterations in genes that covary with some aspects of the body, brain, and behavior, but these features will always be couched with cultural evolution and will emerge and often persist through epigenetic mechanisms.

Finally, for all I have learned, probably the most important revelation in my own journey has been the continuing and exhilarating process of realizing how little I really know, and how much there is still to explore.

LESSONS FROM THE GENOME

Arthur Caplan

With Nathan Kunzler

Mapping the Brain

Inspired by the success in mapping the human genome, two significant projects are now underway to map and improve our understanding of the human brain. President Barack Obama's BRAIN initiative (Brain Research through Advancing Innovative Neurotechnologies) and the European Commission's Human Brain Project. Each project is expected to cost about a billion dollars. Both are to be carried out over ten-year spans.

The BRAIN project was, at the time of its initial announcement, explicitly compared to the Human Genome Project. The hope, says the White House, is that the project will lead to a long list of practical applications, including new ways "to treat, prevent, and cure brain disorders like Alzheimer's, schizophrenia, autism, epilepsy, and traumatic brain injury."

Two decades ago similar promises—many not yet delivered—swirled around government-funded efforts to map the human genome. What can our experiences with the genome project tell us, practically—and ethically—about projects to map the brain?

The Nature of the Two Projects

The BRAIN project states that it provides funding to investigators to develop next-generation technologies with the aim of mapping the activity of each of the neurons in the brain. The goal is to develop technologies for monitoring the responses of large populations of neurons with high spatial and temporal resolution. Initially these will primarily be developed in animal models, like flies, fish, and mice, but with an eventual goal of applications in humans, enabling the study of brain processes, thereby allowing a more accurate study of brain processes from thought and memory to pathologies such as Alzheimer's and PTSD. The list of collaborators includes leaders in neuroscience from all over the United States. Close to half of the initial funding for the initiative is to go to DARPA—the Defense Advanced Research Projects Agency. DARPA's current interest appears to be direct brain stimulation technologies (DBS). This means that a lot of money is being spent on the initiative with an eye toward military use, including DBS for various forms of brain trauma, enhancing or recovering mental functioning, memory modification, brain interfaced prosthetics, and accelerating recovery from brain injuries associated with combat.

The European Commission's Human Brain Project (HBP) seeks to gain insight into the function of the human brain and thereby advance research in neuroscience. It is, however, distinct in the approach being taken. The goal of the HBP is to integrate disparate areas of neuroscience research through innovative informatics and other modalities to create a functional brain simulation. The hope is that functional simulation theories about brain health and pathology can be formulated and tested. The HBP carries less explicit rhetoric about treatments and cures associated with its promotion than does the BRAIN Initiative.

The projects, despite their differences, could prove complementary. Certainly having a neuron-by-neuron map of the brain temporally and spatially would greatly contribute to the ability to create a functional simulation of the human brain. Similarly, information gleaned from simulation using a dynamic model may be key to creating a unifying neuroscientific foundation to guide future research. But there are challenges too, especially when considering the use of new technologies to study the human brain, as opposed to model systems, and it is instructive to consider the history of the genome project to better understand them.

Challenges in Mapping the Human Genome

The first official funding for the Human Genome Project originated with a proposal from then-President Ronald Reagan in his 1987 budget submission to the Congress. It subsequently passed both houses. The project was planned for fifteen years.

In 1990 the two major funding agencies, the Department of Energy and the National Institutes of Health, developed a memorandum of understanding in order to coordinate their mapping efforts. They reset the clock for the initiation of the project to 1990.

Due in part to the prevailing political climate of the United States, which favored private solutions to large-scale projects, there was interest in privately funded alternatives to the HGP. Many felt that private companies dedicated to mapping would find ways to more efficiently and affordably sequence the human genome. Some felt that a project to simply map the human genome was better suited for private enterprise, leaving government funds available for more basic research purposes. Celera Genomics, among other companies, was created in 1998 in partnership with PerkinElmer to perform the mapping work and to profit commercially from the result. Celera quickly became the major competitor to the publicly funded project. The company claimed to be able to achieve the same goals of the project on a faster timetable with a much smaller total budget.

Investors believed they would succeed. Celera Genomics Group stock rocketed after the company, based in Rockville, Maryland, announced in 2000 that it had mapped 90 percent of the human genome. Celera, which began trading at $25 a share, saw its stock price rise to over $200 a share, giving the company at one point a market value of $5.5 billion. Celera's revenue from the sale of genomic sequence information peaked at $121 million in June of 2002.

An issue that came up right away was: who owns the information contained in the genome. Many argued that all genomic information should be publicly available. There was an initial agreement between the public and private groups to share data. This fell apart when Celera refused to deposit their data into a public database—Genbank. This led to a situation where the private project was able to use the data from the public HGP, but the same was not true for the public project in seeking to access the data assembled by Celera, a private, commercial entity. There were no legal grounds for insisting on symmetry.

Meanwhile, the public and private efforts had distinct and different goals. The publicly funded group sought to make human genome information freely available to all scientists across the world in the hope that they would use the information to further the research they were doing in diverse areas of science and medicine. Those in the privately funded group may have had some desire to share information, but (as shown by the hundreds of patents they filed), Celera initially had a strong desire to retain a good deal of the information it discovered as proprietary. Celera was a commercial entity supported by private investors eager for a return on their money. The notion of sharing data was not one that found any support in the company's early days. Only when it became clear that a vague, low-resolution general map of the human genome had no real commercial value did the company move toward a position of freely releasing that map for public use.

Lessons for Brain Mapping

There is enormous value in biological information, whether composing rough low-resoluition brain maps or subsequently fine-tuning them as more precise information is learned about small individual brain variations. High resolution maps of human genomes will have the greatest value in personalized diagnosis or therapies, including creating drug targets. One key lesson learned from mapping the genome is that access to a rough initial map proved crucial to developing more detailed maps of small individual human differences. Unless ALL data, not just crude initial brain mapping data, is guaranteed to be open and freely available, commercial interests and motivations will, as they have in genomics, drive the evolution of knowledge about the brain. While efforts to map the brain have begun as public, government-funded projects, this does not mean that private entities will not enter the arena and seek to compete with those projects.

Although initial efforts to map the brain may be fueled by public funds, the issue of how "fine-tuned" information that can be used to determine risk factors or emerging disease states in individual's brains, which will require linking data to genetic databases, health records, and health databases, will be handled merits discussion now. What rules will govern the sharing of detailed scans or maps about each individual's brain? Can data be linked from a brain scan to a genome to a database without an individual's express consent if that person's identity is not 100 percent secure?

What information about the brain can be patented? Recent battles over the patenting of *BRCA* genomic information by private firms show what can happen if these issues are not acknowledged and resolved early on. It is important to keep new advances in neuroscience from bogging down in fights over commercialization and ownership. Such issues need to be resolved sooner rather than later.

Consider a company formed with the promise of offering customers interesting information about their thoughts and/or predictive information about brain diseases they might be at risk of acquiring. Many such companies, some more legitimate than others, are operating now in the sphere of genomics. Some are huge and have proven profitable, like deCODE and 23andMe. Others are small and often make claims that are on the fringes of genomic science. Building on preliminary and incomplete information coming out of the brain mapping projects and related research, we can predict with certainty that new "brain diagnostic," "truth assessment," and "brain detective companies" will begin to proliferate on the web and elsewhere. The emergence of companies that purport to be able to conduct neuromarketing without much in the way of evidence to ground their claims shows what is likely to be in store in short order regarding "truth" analyses.

All these soon-to-come companies need is some form of a scanner, a suspicious spouse or wary potential employer, and a lot of hocus-pocus to say that new knowledge of the brain will permit the detection of adultery, unfaithfulness, unhappiness, or a disposition to theft. Without any control over the use of new information about the brain or advertising claims allegedly based upon knowledge derived from the new projects to map the brain, the projects will create many spin-offs. Not only will there be spin-off information about how to diagnose disease and treat it and what price ought be charged for such benefits of government research, but there will also spin off a host of quacks, charlatans, entrepreneurs, quick-buck artists, and shysters eager to parlay incomplete or rough data about the brain for sale to a public eager to believe in truth machines, windows into one's deepest hidden thoughts and fears, and screens that can weed out the different, the potentially derelict, and the defective in the home, workplace, or jail.

Calls to map the human genome anticipated none of the aggressive commercial exploitation that followed in its wake. There is no reason not to better prepare for the fallout that will surely occur as knowledge of the brain advances.

What Is the Best Source of Funding a Brain Map?

As was seen in the HGP there are distinct advantages to various types of funding. By allowing public or governmental funding many argue that scientists are allowed academic freedom, an ability to proceed in the direction they feel is most promising. But that belief may be naive. The heavy presence of DARPA, the Defense Advanced Research Projects Agency of the US Department of Defense, in the American project all but guarantees that that project will be under pressure to show benefits useful for national security, military application, and the diagnosis and treatment of combat- and military-service-related disabilities and injuries. There appears to be no guarantee that all data collected under DARPA's auspices will be publicly available. DARPA sponsorship may entice brain scientists eager for grant money in a time of tight budgets, but it is important to realize the goals of DARPA may not always overlap the values of scientists used to relying on NIH or NSF support.

When an endeavor like mapping the brain is funded by private industry funds, or even foundation grants, there is a pressure to move in the direction those funders want. A grant from the Alzheimer's foundation is likely to come with strings attached about mapping with an eye toward better understanding Alzheimer's. The same can be said of industry funds. While they can be a valuable source of money for investigators, they also come with a pressure to find commercializable opportunities in the research.

Foundation and industry grants are not without their merits. The idea that they could lead to more efficient and affordable technologies was used to justify the competition between public and private groups that occurred during the HGP project. The fact is that the competition between public and private efforts to map the genome did in fact lead to more affordable and efficient technologies is now very much appreciated by many in the scientific community. However, if brain projects are fully or partially funded through collaboration with industry to reap the benefits of increased efficiency and translation to application, then it must also be understood that the community of scientists might not be used to working with large corporations, such as GE, Medtronic, Siemens, Johnson and Johnson, Google, and others, and will surely be averse to the demands made on them in terms of proprietary rights as a price for their funding. Large foundations can also make demands that readily create conflicts of interest for those seeking rapid publication and the release of all crude data into public data banks.

Are We There Yet? What Counts as Progress?

Other lessons from the initiative to map the human genome deserve attention as well. The competing projects engaged in mapping the genome did not agree on what would constitute the finish line in terms of announcing success regarding achieving a map of the human genome. Nor did they agree on whose genome or genomes would serve as the template for mapping activities. Often what counted as progress was fiercely debated in public with an eye toward gaining a PR advantage for one side or the other albeit at a serious cost for the public's understanding of what was taking place.

Nor was there agreement on what to map. For example, at the time the HGP was first launched, it was widely assumed that noncoding DNA was "junk" and need not to be taken into account as part of a claim to have "mapped" the human genome. And initially many involved in mapping said that noncoding DNA need not be mapped. But years later, researchers began to realize noncoding DNA played a key regulatory role governing much of the process of epigenesis

When it comes to the human brain, what should we map? Is it a map of the neural connections of the brain, a so-called human connectome? Should the glial matter that makes up as much as 90 percent of the cells in the human brain be included in any map before success is declared? This is an especially important question, looking back on the decision to not include "junk" DNA as part of the human genome. More and more evidence mounts that glial cells are not simply "supporters" of the neurons, as neurological dogma has held for many years, but are possibly involved in brain processes.

As some have argued, building a connectome is not enough to say the brain has been mapped. We also need to develop technologies that can image the brain dynamically and see what cells and groups of cells are firing with high spatial and temporal resolutions to say anything like a brain map has been achieved.

But what resolution will be sufficient? Spatially, should scientists seek to see individual neurons, groups of 5, areas of 1 mm? What about temporally, do we need to see every second, millisecond, or every evoked action potential?

These questions will undoubtedly be the source of much debate among scientists, but they should begin to be addressed now. Without a consensus on what mapping the brain means, at what resolution it will, and ultimately ought to, be done will not be evident, as it was not when seeking to map the human genome. Battles over credit, ownership, error, and the fulfillment of promises of applicability hinge on reaching an agreement about what the endpoints are. Just as importantly, public support and funding for mapping will pivot on clarity about what endpoints are important and what landmarks along the way have real significance.

The Practical Value of a Map

Those seeking to fund the project two decades ago heralded sequencing and mapping the genome as the way to a very rosy future in which we would secure freedom from all our genetic ailments, the key to a longer, healthier, happier life. Indeed, at the official announcement of its completion, then-President Bill Clinton said it would "revolutionize the diagnosis, prevention and treatment of most, if not all, human diseases."

But while genomic technology may well accomplish these things, it is important to recognize that the true advances regarding the "prevention and treatment" of most human diseases are still decades away. Even now, almost fifteen years after the initial announcement of the completion of the project in 2000, medicine is only just beginning to see technologies that may meaningfully change the way human diseases are diagnosed and treated. The public or Congress or other funders might well feel that science did not deliver on the big promises made in the name of mapping or that the time frame that was sold was far too optimistic.

Any large-scale project seeking significant public funds risks facing the same problem as, say, building a location for the Olympics such as Sochi, Russia, or constructing a tokamak for nuclear fusion. Mustering public support for any science project that requires billions of dollars, and, at least in the US context, requires persuading a wary Congress, public, and media that wondrous advances in the human condition lie just around the corner if science can only get enough money, requires reasonable achievable goals, not science-fiction-inspired promises. In order not to disappoint the taxpaying public it is important to be wary of the tendency to overpromise in the name of securing funding. Being able to map the brain of a mouse is no more a promise of cures around the corner than is the capacity to map the genome of a mouse.

Whose Brain or Genome Shall We Map?

At the time the HGP was announced there was a great amount of time spent discussing *whose* genome would be mapped and what the consequences of that decision would be. Because the project would be made available freely to the public anyone who allowed their genome to be sequenced would have to understand and accept the idea that their genetic information would be available for all to see. Any genetic privacy they had would be erased.

In the end a decision was made to sequence the genomes of several volunteers but only after a rigorous informed consent process. The Human Genome Project used protocols to ensure that the DNA from several different volunteers was used and that the blood samples from which the DNA would be extracted were de-identified to the researchers using them. Additionally, many more volunteers were recruited than were needed to sequence the human genome, and as such, no single volunteer is actually certain if their DNA is a part of the project or not. While this approach may have worked well to avoid some of the ethical conundrums of genomic sequencing, it may not be as simple as we map the brain.

The question of whose brain to map is centrally important in the current project for reasons both symbolic and scientific. While many millions of persons have "maps" done of their brains every year for diagnosis or research, at some point a decision must be made about which brain or brains to use as the foundation for a standard brain map. Will we include data from the mentally ill? Will the developmentally disabled be a part of the data pool? Will those who have specific brain diseases be a part of what is brought forward as a "normal" or "typical" human brain, and if not, why not? It seems best to use a group of people that represents a broad section of humanity. This seems to give the best scientific chance of capturing all the important information being sought. It is also, as the HGP found, ethically less cumbersome that choosing the brain of a single, identifiable individual.

This approach, however, may not work. We don't know how different the connectome of each human brain is, and we do not know what sort of variability to expect in a dynamic brain image. It may well be that this variability makes the collected data impossible to pool and de-identify. If brain variability may create issues of brain selection in studying the brain then these issues, which are hugely controversial if they do exist, require public discussion and debate.

Translating New Knowledge of the Brain Will Not Be Easy

No shortage of enthusiasm greeted the first findings to emerge from the crude map of the human genome. Media stories erupted with the promise that "genealyzers" would soon be present in every doctor's office. The Internet also erupted with a parade of scams and nonsensical offerings: genetic testing for predicting athletic performance in children, the best diet suited to a person's genome, ancestry testing, and even the identification of a person's best romantic partner through DNA analysis. So far, little of this has yet emerged from efforts to map the brain. How can we keep brain knowledge from spawning the same sort of hype, confusion, exploitation, and misunderstanding?

Even on some basic concepts, there is already considerable confusion in the general public. Consider the basic concept of brain death—the total and irreversible loss of all brain function—and the recent case of a thirteen-year-old girl, Jahi McMath, who died on December 12, 2013. Her parents had taken her to Oakland Children's Hospital for surgery to remove her tonsils to help her sleep apnea. Things went tragically wrong (although exactly why is not known). She suffered severe bleeding, a heart attack, and massive hemorrhaging in her brain. Unfortunately, experts in neurology could not find any sign of brain activity after these events. Independent experts who were not treating Jahi did the standard accepted scans and tests to assess brain activity and concluded with certainty that she was brain dead.

Yet months later, the girl remained on a ventilator receiving food through a tube in an unidentified facility—because her parents refused to accept her death because they did not accept "brain death." Unlike those in a coma or in a permanent vegetative state like Terri Schiavo, a Florida woman whose parents fought unsuccessfully with her husband to keep her alive, or Ariel Sharon, the former Israeli prime minister whose family kept him in a coma for eight years, no one recovers from brain death. Brain death is death because the brain can no longer support any key vital functions. Of course no parent would want to accept their daughter's death, but because the public continues to confuse brain death with coma or vegetative state, the McMath family received great support from other families and in the media. Indeed, intensive care units in the United States and other nations sometimes contain bodies that have been pronounced brain dead on machines providing artificial life support at the direction of families who cannot or will not understand brain death.

Brain death is widely misunderstood around the world. A brain map is likely to be misunderstood as well unless great care is always used in explaining the concept.

More broadly, if the genome has taught us anything, it's those working to map out biology, be it genome or brain, have a huge social responsibility. The push to map the brain can't just be about gathering information and discussing ways that information might be applied. Scientists must also debunk hype, allay groundless fears, and anticipate likely ways in which efforts may be made to exploit or dupe the public in the name of knowledge derived from brain maps, studies, and scans.

THE COMPUTATIONAL BRAIN

Gary Marcus

Neuroscience today is collection of facts, rather than ideas; what is missing is connective tissue. We know (or think we know) roughly what neurons do, and that they communicate with one another, but not what they are communicating. We know the identities of many of the molecules inside individual neurons and what they do. We know from neuroanatomy that there are many repeated structures (motifs) throughout the neocortex. Yet we know almost nothing about what those motifs are for, or how they work together to support complex real-world behavior. The truth is that we are still at a loss to explain how the brain does all but the most elementary things. We simply do not understand how the pieces fit together.

In my view progress has been stymied in part by an oft-repeated canard—that the brain is not "a computer"—and in part by too slavish a devotion to the one really good idea that neuroscience has had thus far, which is the idea that cascades of simple low level "feature detectors" tuned to perceptual properties, like differences in luminance and orientation, percolate upward in a hierarchy, toward progressively more abstract elements such as lines, letters, and faces. In this chapter, I argue that it is important for neuroscience to break free from both of these ideas.

· · · · · ·

The biggest direct challenge to the notion that the brain is a computer has probably come from Parallel Distributed Processing (PDP) or "neural networks," an approach that dominated the cognitive sciences for almost twenty years, starting in the mid-1980s. In part PDP dominated because it was the first serious alternative to the then-dominant paradigm of understanding intelligence in terms of stored computerlike programs; even then, in the mid-1980s, the mind-as-computer metaphors seemed dated. "Good old-fashioned artificial intelligence," itself modeled on computer programs, seemed to be on its way out; when neural networks ascended in the mid-1980s, a great many people spoke of "paradigm shifts." For the next two decades, there was a palpable sense that a revolution had come.

Then, like so many fads in psychology (Freud's psychodynamic theory and Skinner's behaviorism), neural networks begin to fade away, never quite making the transition from proofs of concept on toy problems (which were abundant) to realistic models of mind or brain. In the 1990s, journals and conferences were filled with demonstrations that showed how it was supposedly possible to capture simple cognitive and linguistic phenomena in any number of fields (such as models of how children acquired English past-tense verbs). But as Steven Pinker and I showed, the details were rarely correct empirically; more than that, nobody was ever able to turn a neural network into a functioning system for understanding language. Today neural networks have finally found a valuable home—in machine learning, especially in speech recognition and image classification, due in part to innovative work by researchers such as Geoff Hinton and Yann LeCun. But the utility of neural networks as models of mind and brain remains marginal, useful, perhaps, in aspects of low-level perception but of limited utility in explaining more complex, higher-level cognition.

Why is the scope of neural networks so limited if the brain itself is so obviously a neural network? Much rests in what is meant by a "neural network." Although the brain is clearly *some* kind of a neural network, the brain is vastly more complex than the *particular* kinds of neural network that people were so excited about in the 1990s. Such networks consisted of simple arrays of input units, output units, and "hidden units." The name of the game was "supervised learning": training a network on some set of known patterns through gradual adjustment of weights so that over time the error decreases; any finite set of examples can eventually be memorized.

Such networks were, as intended, a long way from anything recognizable as computer. They had no recognizable analog to the kind of instructions computer programs are made of (REPEAT . . . UNTIL, IF . . . THEN . . .) or to the variables (x's and y's) that litter virtually any line of computer code. But in retrospect, PDP networks were *too* far away from such things; they were right to emphasize the brain's parallelism but wrong to throw away the computational baby along with the serial bathwater.

Today, "neural network" models have become more sophisticated, but only in narrow ways; they have more layers and better learning algorithms, but they still remain too unstructured. The technique known as deep learning offers innovative ideas about how *un*supervised systems can form categories for themselves, but it still yields little insight into higher level cognition, like language, planning, and abstract reasoning. If there is no reason to believe that the essence of human cognition is step-by-step sequential computation in a serial computer with a stored program (à la von Neumann's ubiquitous computer architecture), there is also no reason to dismiss computation itself. In fact, if I can make a bold claim, I don't think we will ever understand the brain until we understand what *kind* of computer it is.

• • • • • •

One of the biggest mistakes of the neural network movement was in assuming that the brain was initially organized *randomly*, tuned entirely by experience, with no initial systematicity. In reality, there is every reason to believe the biological process of embryological development is capable of building complex, intricate rough drafts of the brain even in the absence of experience, and every reason to believe that detailed circuit structure is critical to nervous system development. To take one powerful example, consider a study performed by the Nobel laureate Thomas Südhof, best known for studying the molecular basis of synaptic transmission. As part of that work, Südhof developed a knockout mouse in which trans-synaptic neurotransmitter secretion was genetically silenced, thereby shutting down most of the brain's internal communication, hence, much of the ability of those mice to learn. If the initial drafts of the brain were organized primarily by experience, you might expect the knockout mice to have essentially random brains at birth. Instead, to a first approximation, by birth the synaptically silent mouse embryos had developed brains that appeared to be more or less normal, with the folds, the gyri, the different cell types, the regularly organized structures, and so forth that one would expect. Subsequent work by Zoltan Molnár (in physiology) and Giorgio Vallortigara and Lucia Regolin (in behavior) pointed in the same direction: much of the brain's basic organization can be structured even in advance of experience. Experience tunes, calibrates, reshapes, and rewires, but it is only half the equation.

Going hand in hand with the neural network community's odd presumption of initial randomness was a needless commitment to extreme simplicity, exemplified by models that almost invariably included only a single neuronal type, abstracted from the details of biology. We now know that there are hundreds of different kinds of neurons, and the exact details—of where synapses are placed, of what kinds of neurons are interconnected where—make an enormous difference. Just in the retina (itself a part of the brain), there are roughly twenty different types of ganglion cells; there, the idea that you could adequately capture what's going on with a single kind of neuron is absurd. Across the brain as a whole, there are hundreds of *different* types of neurons, perhaps more than a thousand, and it is doubtful that evolution would sustain such diversity if each type of neuron were essentially doing the same type of thing.

• • • • • •

There are, of course, many reasons to think that brains operate mostly in parallel. Individual neurons are too slow to allow brains to operate in strict serial von Neumann fashion, and ample data suggest that in any given laboratory task (and by extension, any real-world situation) many different parts of the brain are engaged simultaneously. Even when we are not involved in any specific task, a so-called default or resting state network fires away, with many different neural circuits operating in parallel. But that in itself doesn't militate against the idea that the brain might be some kind of computer; computers are often portrayed as if they were invariable serial sequential devices, but the reality is that for the last twenty-five years, since personal computers became popular, there has always been some degree of parallelism: an input-output controller working alongside the central processing unit, for instance. By the 1990s Graphics Processing Units (GPUs) started to become popular, acting as coprocessors with a central processor, taking up most of the work of displaying images so that the CPU would be free for a program's main logic. And importantly, GPUs were themselves computers, but ones with a dedicated job—essentially matrix arithmetic—and one that they did almost entirely in parallel. Later, "multicore" processors came to be popular. Modern computers (and for that matter smart phones) are by any reasonable measure computers—systems that manipulate information systematically—but not at all strict von Neumann machines with a single stored program executed in purely sequential fashion. The idea that brains can't be computers because computers aren't parallel is mired in a vision of computers that is thirty years out of date.

What GPUs and CPUs have in common is that each revolves around a basic set of *instructions*, such as addition, subtraction, and multiplication, that work in "algebraic" fashion, such that those operations can work over arbitrary values that are stored in a set of registers. A kind of stereotypical process in a GPU, for instance, is to darken every pixel (a form of subtraction) in an image by a fixed amount simultaneously. A classical serial computer may accomplish the same thing pixel by pixel or byte by byte, but the results would be the same. Once one realizes what a GPU can do, and realizes that a GPU is just a different kind of computer, the notion that the brain might somehow *not* be a computer loses all its force. Many pathways in the visual cortex, for instance, seem to perform transformations on representations of visual scenes, for example, extracting, in parallel edges across a scene. Digital designs like ASICs that are dedicated to specific tasks (like BitCoin mining) show that programs are optional, too; up to certain limits, many programs that might be loaded into memory and executed sequentially can be translated into parallel circuitry that is hardwired and run without a stored program.

In my own view, it is obvious that brains (especially those of vertebrates) *are* computers, in the sense of being systems that operate over inputs and manipulate information systematically. Brains might not be (purely) *digital* computers, their memories may operate under different principles, and they may perform different sorts of operations on the information they encode, but they surely encode information. For example, by transducing inputs into patterns of chemical and electrical information, they manifestly operate over that encoded information, and they use the resulting outputs to do things like guiding motor action and updating internal representations. Computers are, in a nutshell, systematic architectures that take inputs, encode and manipulate information, and transform their inputs into outputs. Brains are, so far as we can tell, exactly that.

The real question isn't whether the brain is an information processor, per se, but rather *how* do brains store and encode information, and what *operations* do they perform over that information, once it is encoded. The mission of neuroscience, in my view, thus ought to be to reverse engineer the brain in much the way that one that might try to reverse engineer a GPU. An investigation of a GPU would initially reveal that the basic elements are transistors, and eventually it would become clear that those transistors were organized to execute a relatively small number of "instructions," such as lightening an image or rendering a polygon. More complex processes would be amalgamations of those instructions. In our understanding of the brain, we recognize that neurons are the analogues of transistors, but we know too little about the operations of those individual neurons, how they encode information, and especially how they manipulate that information. In computers, our understanding of how information is manipulated begins at the circuit level, at which transistors are assembled into circuit motifs, to create basic logical operations (or "primitives") like AND, OR, and NOT. A circuit-level understanding of neurocomputational *primitives* is likely to be fundamental for decoding the brain as well.

Technologies for measuring activity and connectivity across ensembles of neurons, many described in this book, clearly give us some of the tools we need to begin to develop that understanding. Even so, what I still believe to be lacking is a *theory* about how sets of neurons might come together to support something as complex as human cognition. Neuroscientists often encourage each other to work in a strictly bottom-up fashion, reviewing known facts from physiology while scarcely paying attention to more abstract hints from behavior and from computation. Epitomizing this view, some less enlightened neuroscientists have been known to say, "data talks, and theory walks"—a perspective that, at least in my view, has impeded neuroscience's progress; theory, which should be indispensable, has become marginalized at best. (Theorists don't always help matters, since many theorists seem to seek confirmation that their own particular account is correct; too few seek to compare plausible alternatives.)

Will more data alone be enough to solve the problem of understanding the brain? I doubt it, not in itself; what will really solve the problem is a framework for understanding how the brain might *even in principle* do what it does. And that means, first and foremost, figuring out what kind of a computer the brain might be, formulating competing hypotheses, and testing them, not collecting data first and asking questions later.

• • • • • •

The best hypothesis of how the part of the brain that is unique to mammals—the neocortex—works is that it is a hierarchical array of feature detectors, proceeding from bottom-up sensory information to higher-level more abstract concepts; low-level detectors perceive elements such as edges and curvature, which in turn are fed into nodes that detect complex stimuli such as letters or faces. This idea goes back to Hubel and Wiesel's work, and—to a certain extent—it's almost certainly true. Many neurons specialize in detecting low-level properties of images, and some neurons that are further up the chain of command represent more abstract entities, like faces versus houses, and in some instances, even particular individuals (most notoriously, Jennifer Aniston, in work by Itzhak Fried, Christof Koch, and their collaborators). The "Aniston" cells even seem to respond cross-modally, responding to written words as well as to photographs. Hierarchies of feature detectors have now also found practical application, in the modern-day neural networks that I mentioned earlier, in speech recognition and image classification. So-called deep learning, for example, is a successful machine-learning variation on the theme of hierarchical feature detection, using many layers of feature detectors.

But just because *some* of the brain is composed of feature detectors doesn't mean that *all* of it is. Some of what the brain does can't be captured well by feature detection; for example, human beings are glorious generalizers. In my own lab, for example, we found that seven-month-old infants could pick up on the regularities in strings of sentences made up according to an abstract grammar. In just two minutes' exposure to a set of sentences following an ABB grammar like *ga na na*; *la di di*, babies learned a rule that they could generalize to new words, distinguishing between synthetic "sentences" like *wo fe fe* (that followed the same grammar that they had been exposed to) and *wo wo fe* (that followed a different grammar, in this case of the structure AAB). A more recent study using brain imaging replicated this result, but in *newborns—* suggesting that the capacity to detect such abstractions may well be innate. Hierarchies of feature detectors can learn well to *classify* anything that you've seen before repeatedly, the more often the better, but they still lag behind babies in extending abstract inferences to new cases.

Hierarchies of features are less suited to challenges such as language, inference, and high-level planning. For example, as Noam Chomsky famously pointed out, language is filled with sentences you haven't seen before. Pure classifier systems don't know what to do with such sentences. The talent of feature detectors—in identifying which member of some category something belongs to—doesn't translate into understanding novel sentences, in which each sentence has its own unique meaning.

At its core, language revolves around a process known as variable binding. For instance, simplifying a bit, you might say that English has a rule that says that a SENTENCE can be composed of a NOUN PHRASE and VERB PHRASE, where the variables are in SMALL CAPS, and can be filled in a potentially infinite number of ways, yielding a potentially infinite range of meanings. The beauty of a rule like that is that it is potentially infinite, encompassing everything from *the sailor* (noun phrase) *loved the girl* (verb phrase) to *the beauty of a rule like that* (noun phrase) *is that it can encompass virtually everything one might wish to say, even if it has never been said before* (verb phrase). To explain what Chomsky has called *discrete infinity*, we will need something beyond hierarchies of feature detectors, which specialize in classifying what we have seen before but lag in allowing us to interpret things that are new. To understand the neural basis of human cognition, we will need to understand, in particular, what linguists call *compositionality*: the way in which our brain allows us to put together smaller elements (like words) into larger, straightforwardly interpretable complexes (like sentences), even when those larger complexes are novel.

• • • • • •

To close this deliberately provocative and opinionated piece, I would like to pose six specific challenges, or questions, difficult but not insuperable, in hopes that progress in any one might move the field ahead significantly.

1. If the brain is not a von Neumann stored program machine in which software is loaded into memory and followed in step-by-step

fashion (as von Neumann himself recognized), what kind of an information processor is it? How does the brain manage to be so coordinated in the absence of a central clock? Is there a kind of neuronal algebra, a set of operations that work on arbitrary values stored in synapses? (Or, for those who would doubt that brains *are* computers, is there any serious alternative?) To the extent that the *human* brain is capable of algebraic computation, as I have argued, are we alone? Can other mammals or other vertebrates perform similar operation?

2. Although the human brain may occasionally approach von Neumann–style computing—as in conscious, deliberate rule application (for example, what a beginning trigonometry student does, based on verbal instruction), most of what it does probably shouldn't be characterized in that way. What computations do we use in other domains, where knowledge and instruction are both less explicit? What kinds of neural systems could conceivably support the versatility of our cognition? Remarkably, we still haven't even resolved the basic question of whether brains are analog, digital, or (as I suspect but certainly can't prove) a hybrid of the two. (Brains might, for instance, use digital computation for grammar and analog computation for some aspects of image processing.)

3. How does the brain implement variable binding, and what sorts of operations can a brain perform with variables once they are bound? Variable binding, akin to setting x to equal the value of 5 in order to calculate an algebraic equation such as $y = x + 2$ is a central process, possibly with multiple realizations, that arises at many levels. Variable binding is central whether we are tracking moving objects (this animal I am chasing, as opposed to that one, even though they look identical) or putting together the elements of a sentence. In a sentence, for example, the variable noun phrase must be temporarily associated—bound—with a particular string of words (for example, "the man who went up a hill but came down a mountain"). How do such temporary relationships get established? The standard accounts of memory depend on hundreds of conditioning trials, yet we establish dozens of short-term

bindings every time we comprehend a single sentence. Nobody knows how the brain does this.

4. Is there a single canonical form of computation (like hierarchical feature detection), as is often assumed, or a wide range of basic operations, recruited over and over again, much like the instructions in a microprocessor, as I am arguing? If it is the latter, what is the range of basic operations, and how are they realized, neurally?

5. What format(s) does the brain use to encode information? Computers use encoding schemes like the ASCII code for letters, JPEG and GIF for images, and so forth. How does the brain encode a sentence? A word? A mental image? A melody? We have some hints as to how the brain encodes targets in motor space (see chapter by Shenoy), and how it represents Euclidean space (see chapter by Moser and Moser) but we know desperately little about the brain's other formats for representation.

6. Why does the brain contain so much diversity, at every level of analysis? From the 100+ cortical areas in the human brain, with vast numbers of apparently orderly connections between them, to the hundreds of neuronal types, to the enormous amount of molecular complexity within individual cells and synapses, the dominant theme of the brain is not simplicity (as so many computational neuroscientists seem to hope) but complexity; crucially, too, the brain is a delicate system. Although an ordinary brain can develop in a wide range of environments, neural disorder consistently tends to be associated with mental disorder. If the right cell types aren't connected in the right way, mental illness or mental retardation is often the result. There is an enormous amount of detail in the brain, and the details seem to matter, a lot. What is all the detail for? What do you get from a complex and diverse brain that would not emerge from a large but simple neural network?

Acknowledgments: Ned Block, Jeremy Freeman, Christof Koch, and Athena Vouloumanos provided superb comments; all remaining hyperbole is my responsibility alone.

Further Reading

Marcus, G. F. 2001. *The Algebraic Mind: Integrating Connectionism and Cognitive Science*. Cambridge, MA: MIT Press.

Marr, D. 1982. *Vision: A Computational Investigation into the Human Representation and Processing of Visual Information*. Cambridge, MA: MIT Press.

Poeppel, D. 2012. "The Maps Problem and the Mapping Problem: Two Challenges for a Cognitive Neuroscience of Speech and Language." *Cognitive Neuropsychology* 29 (1–2): 34–55.

IMPLICATIONS

As the most complex system in the known universe, the human brain would merit study even if there were no immediate practical significance; the fact that a three-pound piece of meat can perform many intellectual feats that still surpass our greatest computers naturally makes the brain an object of wonder. But the brain is also the last great target in medicine; the more we can understand it, the more we will be able to shape our destinies. What happens when we do understand the brain?

John Donoghue describes what might be accomplished—medically—once we have a better understanding of the brain, and how the line between man and machine might start to blur. **Kevin Mitchell** describes the challenges in treating psychiatric disorders, and how a better understanding of the genetic contributions to the brain might break the logjam and lead to much more effective treatment than is currently possible. **Michel Maharbiz** describes future technologies that might eventually allow us to obtain, in humans, the same kind of detailed neural recordings that are currently only available in nonhuman animals.

NEUROTECHNOLOGY

John Donoghue

We currently lack a deep understanding of how brains operate and how a brain's obscure operations produce behavior, especially those behaviors that are highly specialized in humans. Superficially, brains represent and store activity patterns, and then, sometimes, transform them by "neural computations" to generate an overt behavior. How the collective actions of neurons embedded in immensely complex circuits "represent" and "compute" has been elusive. This ignorance also profoundly limits our ability to treat many of the most debilitating brain disorders, like depression, autism, epilepsy, schizophrenia, or paralysis, which emerge from impaired circuit function. But some of those deficiencies may soon be overcome because of the ongoing neurotechnology revolution, which will be disruptive in three spheres. First, new tools will provide a means to comprehend the basic principles that link brain activity to core mental functions of humans—perception, cognition, emotion, and action—and, for the first time, explain mechanistically unique features of human brains. Second, this technology will yield a new class of "brain interfaces," which are likely to transform the way clinicians interpret and treat brain disorders and especially provide a physical way to restore lost functions. Lastly, neurotechnology advances may ultimately challenge our views of what it means to be human. Below, I consider the impact of neurotechnology in each of these three spheres.

Finding the Middle Ground: New Tools, New Rules

Networks of neurons, working together in intricate circuits, can produce complex functions like emotion, cognition, or planned behavior. In the quest to understand how these properties can emerge from the collective action of large numbers of neurons, perhaps the biggest knowedge gap is to link function across scales. At the lowest level of function, the neuron, we have a reasonable, although certainly incomplete, understanding of how individual neurons transform input to output. We have connectional wiring maps at varying degrees of resolution in many species, even man. And at the highest levels, thanks to tools like MRI machines, we can indirectly infer how global brain activity patterns of millions of neurons collectively engage during thought, emotion, or overt behavior. Neither the single cell nor whole brain level can explain mechanisms that operate at the middle, or *mesoscale*, where behavior emerges from rich cellular interactions.

Brain operations include the collective dynamics of large numbers of spatially distributed neurons working in highly interconnected networks of ever-changing configurations. Mesoscale operations employ circuits that combine memory and percepts to generate plans, which may then be executed as actions—from typing a sentence to performing elegant dances or to speaking profound prose. Alternatively, plans and ideas can be held in memory as long as a lifetime. To provide a functional map that explains how action, cognition, or emotion emerges from collective network dynamics, we will need new tools that combine a dense sampling of spatially distributed neurons with high temporal resolution in a behaving brain. Exactly how large a network or how many cells must be sampled is a big, open question.

Extant tools like the microelectrode, circuit tracers, and molecular labels have provided a massive base of fundamental knowledge. We can chart neural routes from perception to action, especially for visually guided behavior. For example, we have a very precise map of the path from the eye, to the thalamus, to the primary visual cortex in mammals. We know that this circuit then spreads through two cerebral pathways that process what an object is, and where it is. Through connections with the frontal lobe, a vast network can formulate and carry out a plan to reach, grasp, and manipulate almost any object you perceive, or even one you remember but can't see. Single neuron studies have cataloged feature selectivity (parts of the object) at each step, and fMRI has revealed the peaks of activity as each zone engages. But in reality, current tools are not yet adequate to analyze the mesoscale operations entire networks perform. Perceiving and acting lead to concurrent activation of areas working together in broad networks—not individually. These networks connect nearly every neuron to almost every other neuron in just a few links. Some networks have more connections, perhaps forming central hubs, while others have few. How the impact of sparse and dense connections affects global function is one of the great questions at the mesoscale. Adding just a few functionally relevant connections here or there can profoundly change the properties of interconnected groups, as studies of small-world networks revealed—the kind of networks that explain the well-known six degrees of separation/Kevin Bacon phenomenon.

Neurons are unique processors. Unlike all-or-none transistors in digital circuits, most neurons in the brain combine individually weak signals from thousands of other neurons. They collect this information through synapses: the junctions where neurons contact each other. Synaptic influences from large numbers of cells combine on a target neuron to produce an electrical output. This output consists of a signaling train of brief pulses, known as action potentials, or more colloquially as "spikes." Information is conveyed by changes in the rate of spiking. To add to the complexity of circuits, connections between neurons can be shaped by experience to change their strength on short notice, using a feature known as synaptic plasticity. Synaptic influences can further restructure circuit actions as neurons are bathed in combinations of chemicals called neuromodulators. Information being processed in the brain thus appears, at the middle scale, as changing patterns of spiking in neuron networks, networks comprised both of local circuits and expansive ones.

At this level, a mesoscale neural computation can be considered as the transformation of patterns of spiking activity across these "networks of networks" (figure 1). Tools to map collective dynamics at single neuron resolution, over really large numbers of neurons, are inadequate or lacking altogether, but they are coming. Once network function can be mapped, the ability to selectively manipulate circuits is a second essential tool required to test the role of particular network processes in behavior.

First steps to validate the usefulness of middle scale recordings have been achieved. Important operating principles of small neural circuits have been revealed in detailed studies of isolated clusters of neurons of simpler creatures like worms and lobsters, where every part of a

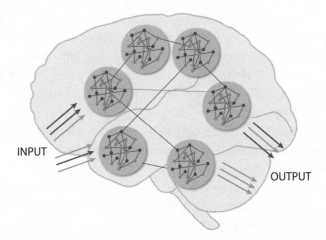

Figure 1. Illustration of a network of networks, to provide a sense of the complexity of understanding brain circuits. Each shaded circle encloses a locally interconnected network of neurons. Some local networks are connected with others, with connectional details not shown. The complete assembly forms a global network that could be thought of as containing a mental state. The input, shaped by the current state dynamics of the whole assembly, "computes" an output pattern (behavior). Note that the effect is shaped by the moment-to-moment influences across the network, but would also be affected by plasticity that shapes the impact of each connection and the precise biophysical properties of each cell. How this is accomplished could be investigated by recording all or a sample of all the elements. This diagram fails to reflect the identify of the cell types, details of local and global connectivity, the time evolving nature of the interactions, the plasticity of the network with experience, and the influence of global neuromodulatory transmitters, among other shortcomings that are important to fully characterize the system. Emerging tools suitable to measure and manipulate activity of sets of neurons will allow hypotheses testing of the nature of representation and computing across the entire circuit.

well-defined circuit can be studied. Scaling up the many orders of magnitude needed to adequately map the collective dynamics of highly specialized brain functions of mammals is not yet possible, but more limited sampling of networks provides clues that this is a fruitful path. The activity pattern of a few dozen cells, which can be now recorded all at once in a local cortical network in rodents, monkeys, and even humans, have revealed mesoscale operations that code the path of spatial navigation in a maze and hand direction when reaching for a goal.

To gain an intuition of how collective neural activity patterns can code behavior, imagine a simple scenario in which we record spiking from two neurons in the motor cortex of an animal as it reaches from one point to another. In the motor cortex, commands to move are assembled through interactions with many other brain structures (that is, a network embedded in a network). Suppose we observe one neuron that spikes significantly more when an animal reaches to the left than to the right. The other neuron spikes maximally when reaching up and very little when reaching down. We now have a coding scheme: neuron one is "tuned" to left–right movement and neuron two is tuned to up–down movement. If we later observe both neurons and find that both are spiking at high rates, we can assume that a reach upward and to the left is being made. Thus this little ensemble of two neurons can be decoded to predict behavior. Reaching direction emerges from the collective behavior of this (minimal) network. While the actual activity of larger assemblies is considerably more complex, the basic idea holds at higher scales. Understanding that neural populations encode this emergent feature of reaching direction in two dimensions gives new insight into how the motor cortex creates movement commands. This knowledge has been exploited to create a brain-computer interface (BCI), where decoded activity patterns from neural ensembles can be read out from the brain of a paralyzed human, allowing them to control computers and robots "at will."

Current reconstructions of behavior from neural activity patterns are not remotely complete because only small samples of massive circuits have been viewed in action. Current multielectrode arrays, with about one hundred sensors, can only measure a very tiny fraction of any meaningful network in a mammalian brain. Ensembles hide many levels of information as they scale up, including features that can combine perception, cognition, motivation, or emotion in their more complex spatial or temporal activity patterns. These features might only emerge when very large numbers of neurons are recorded for long periods of time. Imagine trying to understand a car chase scene if you just saw a random 1 percent of your TV's pixels for a second or two. Technical barriers have limited the ability to (1) fabricate neuron-resolution sensors capable of stably recording large assemblies of cells over long times, (2) build processors capable of processing staggeringly large amounts of information, and (3) use or create the right analytical tools able to extract meaning and mechanism from the activity of very large collections of neurons.

However, advances in neurotechnology, stemming from progress in the physical sciences, engineering, and synthetic biology as well as advances in statistics, mathematics, and computer science, should finally enable an all-out attack on these problems. Capitalizing on nano- and microfabrication techniques, cheaper, smaller, and faster electronics, and huge increases in computing processor and data storage combined with analytical tools to model and simulate patterns from immense numbers of neurons, could finally create the toolbox necessary to understand the mesoscale dynamics in networks of networks in order to link cells, circuits, and behavior.

Neurotechnology for Sensing and Stimulating Circuits

Two avenues for sensing, or reading out, collective activity already show great promise as mesoscale tools: electrical and optical. Direct sensing of a spike's brief, very weak electrical impulse requires nuzzling a hair-thin, fine-tipped microelectrode close to a neuron. Recording many neurons requires many electrodes. It is already possible to insert arrays of up to one hundred microelectrodes into the cortex to study local assemblies of about fifty to one hundred neurons. But scaling up to more neurons is challenging because large numbers of probes could produce unacceptable tissue damage. In addition, obtaining reliable, stable, and long-lasting recordings is difficult because materials degrade and rigid sensors wiggle. Reading signals from hundreds of electrodes has required electronics that are big and bulky with complex plugs and many wires that are prone to failure. This is where engineering can now come to the rescue. Thanks to nano- and microscale fabrication techniques it is now possible to craft electrodes so that they record not just at their tip but at hundreds of recording sites along the electrode's shaft; they can be much smaller and more flexible. And electronics with vast processing power can be reduced to the size of a small matchbox that can be implanted under the skin. Thanks to advances in wireless transmission, impressively large amounts of data can be broadcast out of the body from these implanted devices using radio or light. None of these recent technological advances have succeeded yet in producing dependable electrical recordings for thousands of neurons, but this work is underway and feasible to achieve. Advances in synthetic biology, electronics, and nanofabrication are sparking all types of creative solutions. One far-reaching concept for even broader—and perhaps less intrusive—recording is to make complete packages of electrodes—with integrated electronics and transmission—as small as large specks of dust, so that even thousands might be placed in the brain (see also chapters by Maharbiz and Church for other approaches). Of course, even if fabrication of such technology is possible, it is not yet clear how the brain would react to these aliens, nor is it clear that we can protect or power them.

Light is an alternative to directly recording electrical activity. Genetically encoded indicators can make large numbers of neurons individually report their electrical activity indirectly by emitting light (see chapter by Ahrens). These optical signals (using calcium or voltage indicators) can be detected from surface-mounted and nonpenetrating sensors in freely behaving animals. Optical recordings of up to tens of thousands of neurons have been obtained in a larval zebra fish, and this technique is working in a limited way in small mammals. However, current methods again have spatial and temporal limitations. The field of view from the brain's surface is limited to a few mm square patch and is less than the full depth of the 2 mm thick cortex; even this restricted view requires a hole through the skull. Access to deeper structures requires probes that penetrate the brain. At present light emitting labels are not able to keep up with the speed of spiking, or they can interfere with cell function. Optical reporter delivery systems such as viruses are impressively selective but can be unreliable. Fortunately, there are vigorously active and highly promising initiatives attempting to overcome all of these pitfalls, including the creation of new microscopes small enough to be worn on a mouse's head and new optics that can peer deeper into the brain and read out more quickly.

At the same time, these two approaches—electrical and optical—are being exploited as ways to *manipulate* circuits. Establishing the causal link between circuit function and behavior requires the ability to intervene at specific points in a functioning circuit, not just to observe. Electrical stimulation is actually a very old way to probe neural circuit function. More than 150 years ago the connection of cortical motor areas to movement was tested by evoking a muscle jerk when the brain's surface was electrically shocked with a large electrode. Now, microelectrodes inserted into circuits allow much more precise electrical stimulation to add a signal or disable a signal at critical points. However, even focal electrical stimulation still lacks precision because neural processes the electrical impulse affects are locally entangled nearly everywhere in the brain. By contrast, optogenetic methods provide unprecedented selectivity to turn cells on or off so that the role of selected networks in behavior can be directly tested. Optogenetics promises greater selectivity for manipulating each element of a circuit, but will surely have its own pitfalls. Thus, electrical and optically based neurotechnology are poised to provide the missing middle level of data and key information that links neurons to behavior.

Neurotechnology as a Clinical Tool

Advances in neurotechnology sparked by basic research will also generate an entirely new set of diagnostic, therapeutic, and restorative devices for human clinical applications. Importantly, the knowledge gained from using these same tools for basic inquiry will further enhance clinical use. Understanding principles of circuit function will improve the ability *write in* missing spatial and temporal activity patterns that can be used in devices to replace lost senses or to modulate diseased brain circuits. Better tools to *read out* information hidden in neural activity patterns could reveal the nature of disordered circuit function producing psychiatric disease or restore motor commands from the brain to the body after they were cut off by stroke. Some examples of these applications follow.

Writing In

Cochlear implants—the first wearable clinical neural interfaces—convert sound into electrical impulses and deliver them to auditory nerves in the ear when hair cells—receptors that transduce sound waves into patterns of electrical impulses—are destroyed. This now decades-old FDA-approved technology has provided a profoundly enabling restoration for more than 200,000 people with hearing loss. Cochlear implants demonstrate the power of even rudimentary technology, when coupled to an adaptable brain, to better the human condition. Surprisingly, sound comprehension can be accomplished using fewer than two dozen stimulating sites from a thread-thin probe inserted near neurons in the inner ear, approximating the job ordinarily done by thousands of hair cells. A similar approach is underway to restore vision for those who have lost their photoreceptors from diseases like macular degeneration or retinitis pigmentosa. A video camera can already deliver simple patterned stimulation on a sixty-four-point grid placed against the human retina to engage neurons carrying information to the brain. Crude vision is thus possible. More fine-grained patterned stimulation, guided by better understanding of natural activity patterns and delivered through more sophisticated electronics, could lead to something approaching what we take for granted as normal vision or hearing. A more far-reaching but potential application of optogenetics is to recreate light sensitivity in remaining retinal cells in order to replace missing photoreceptors altogether. Improvements in the spatial and temporal stimulation patterns informed by the mesoscale operations of the retina or cochlea, or their pathways in the brain, will further push performance toward fuller neurorestoration with physical devices that can replace damaged or missing networks.

Rebalancing Circuits for Movement, Mood, and Memory

Neuromodulation is the use of targeted stimulation to adjust circuit activity in brain disorders. Deep brain stimulation (DBS) employs millimeter-scale electrodes to electrically alter neural circuits. DBS systems, already implanted in the brains of more than 100,000 people, reduces the rigidity and tremor of Parkinson's disease (PD). DBS is also being evaluated in clinical trials for depression, cognitive decline, and a wide range of other disorders where circuits somehow malfunction. While DBS can have a life-transforming impact, vague understanding of the network underpinnings of DBS effects, as wells as the disorder itself, is evident in the variability of clinical outcomes, often requiring ongoing adjustment of stimulation patterns and ongoing medication. DBS for Parkinson's intervenes in a network of interconnected cortical and deep structures that have an ongoing loss of the neuromodulatory influence of dopamine. Thus basic research on operations of normal and abnormal function of this circuit should lead to more principled approaches to neuromodulation therapy. In addition, the imprecision of today's relatively large electrodes will benefit from advances in neurotechnology that allow precise, targeted spatial and temporal stimulation. Optical approaches seem to promise even greater potential, if humans can have their neurons rendered light sensitive, because this approach would provide cell-specific selectivity not possible with direct electrical stimulation. Over the longer term, one might envision ways to deliver energy with the appropriate selectivity from outside the head, eliminating the need for surgical placement of electrodes. Magnetic coils, ultrasound, and light have the potential to go through the scalp and skull but are currently too coarse to target specific circuits. Notably, however, noninvasive transcranial magnetic stimulation (TMS) transiently relieves symptoms of depression in some patients despite its inability to be very precise. The poorly understood mechanisms by which TMS affects neural networks should be revealed as the normal operations of these circuits and the effect of disease and stimulation are better understood by the coming wave of new tools. Neuromodulation illustrates how neurotechnology bridges basic research and clinical application. Stimulation in the research setting can inform clinical therapy, and the outcome of clinical use of stimulation can prompt new research questions. Importantly, people who are receiving stimulation as therapy or are participating in clinical trials are providing scientists with a new window into the human brain in health and disease. Human research participants are partners with scientists and clinicians in this new era of research. However, this participation requires careful oversight when cognition or emotion are being manipulated. While consent may be straightforward in diseases such as Parkinson's, where the impact of the disease is predominantly physical, the situation becomes more murky when patients with psychiatric disorders or memory loss are involved.

Reading Out: Turning Thought into Action

Sensing technologies will have a growing clinical impact in the future, especially as innovations in network recording improve. Brain-computer interfaces (BCIs) have attracted particular attention as a sensing neurotechnology with large clinical impact. BCIs attempt to restore lost independence and control for people who are paralyzed. It is a physical nervous system that is a new communication channel to reconnect the brain to the outside world. Through a sensor, signal processors, and computers, the technology detects and decodes neural ensemble activity patterns in order to create replacement commands for actions that cannot be performed. Strokes, spinal cord injury, degenerative diseases like ALS, multiple sclerosis, or limb loss can disconnect the brain from the body (by destroying communication paths) while leaving the brain intact. Each condition prevents intention, formulated in the motor areas of the brain, from becoming a movement. BCIs with even with minimal capability, say to click a switch for yes and no, have great potential value to people so severely paralyzed that they cannot move or communicate in any reliable way. In its most ambitious form, movement commands from the brain could be used to operate machines like computers, prosthetic limbs, or robotic devices, or in its loftiest form, to activate paralyzed muscles to create a physical replacement of missing neural circuits. These ideas, seemingly fanciful, are already in human testing.

BrainGate, a BCI system being developed by our collaborative team, directly connects a part of the cortical arm movement network to assistive technology. In a small group of humans with severe paralysis who are part of early stage clinical trials (currently FDA-limited to investigational use), a baby-aspirin-sized platform with one hundred protruding microelectrodes has been implanted in the arm region of their motor cortex. This multielectrode sensor detects the spiking pattern of a few dozen neurons and passes these signals to external electronics, where a computer algorithm decodes the pattern into useful movement commands. Decoding of this very limited sample of a large network of neurons is surprisingly accurate enough for participants to operate a computer or feed themselves or drink using a robotic arm.

To understand how this apparent magic is possible, recall the earlier discussion of reach coding by two spiking neurons. Spiking patterns, even when imagining arm movement, carry enough direction information to allow pointing a cursor at letters on a screen to type, or moving a robotic arm to reach and grasp. Notably, however, these movements are not as fast, accurate, or dexterous as those a natural human arm performs. *Why not?* Knowledge of how movement intentions are coded in brain networks is not yet sufficient to map activity onto every desired action. For example, we don't know the scale (that is, the size, distribution, and dynamics of the population) required to achieve the flexibility, speed, or dexterity easily accomplished when you use a fork to eat, using an intact nervous system.

BCIs illustrate a path by which a basic research tool to study coding in neural circuits, the multielectrode array, has been co-opted as part of a new human clinical device. The tools and rules coming from mesoscale basic research will lead to better and faster BCI control. Wireless transmission of multichannel neural signal patterns, highly valuable in research to study brain activity during natural behavior, will provide untethered, full-time BCI use for people. A better understanding of sensory coding in cortical circuits could allow a recreation of sensory percepts using patterned stimulation, closing the sensorimotor loop so that touch could guide action. Mesoscale neural decoding has considerable potential for other clinical applications, such as epilepsy. Here, very sensitive measurements of network dynamics could be used to spot aberrant collective activity. In a clinical device, it might be possible to predict seizures well before they have clinical manifestations, and precise, targeted stimulation might be used to abort them.

Next-generation tools promise to expose large-scale network function with high temporal precision across large areas of brain. As we better understand the collective dynamics of neurons in behavior, thought, and disease, new brain interfaces should enhance the ability to restore complex vision, more naturalistic hearing, and dexterous movement; neuromodulatory devices may eventually reduce or even eliminate the manifestations of circuit disorders that produce movement, mood, or cognitive disorders. Not only do these applications have the potential to restore life quality, they also will ultimately reduce the enormous costs of treating and managing people whose independence and quality of life is diminished by these disorders.

No one would reasonably expect complete or immediate success in all of these ambitious clinical applications, nor is it possible to project when they will be fully realized. Advances that reveal the nature of neural collective dynamics will assuredly generate many new questions. Clinical trials may face side effects or device failures that slow progress. Nevertheless, continual gains in understanding network function in humans and animals, and in the tools required for that process, will have a transformative impact on the treatment of neurological and psychiatric disorders.

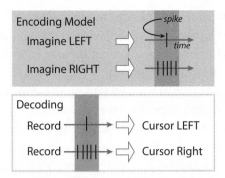

Figure 2. Neural decoding. This simple example explains how commands can be decoded from the pattern of action potentials ("spikes"). In this case, just one neuron in the motor cortex is illustrated. We record this neuron's activity (spiking) while a person imagines moving their hand to the left as they watch a cursor move left on a screen as if they were using a mouse to make that cursor movement. In reality they are not moving the cursor, our computer is moving it to the left automatically. This animated action guides the speed and direction the participant is supposed to imagine. During that time we count the number of spikes that occurs in some defined time window (in blue), which is "1" in this case. Now this same "user imagine and computer count" process is repeated for imagining a rightward cursor motion, which results in "5" spikes in our example. These data allow us to build a model in which 1 means left and 5 means right. In the future, we can DECODE neuron activity and use the observed spike count to drive cursor motion. When we observe 1 spike in our window, the cursor will be moved incrementally to the left, or to the right if 5 spikes are observed. Of course, neurons in the motor cortex don't fire this reliably. Averaging across many neurons allows a better estimate of the intended action, i.e., a better model of what is being intended. More sophisticated mathematical methods can help to improve both the quality of the model and the decoded output, for example, what to do when 3 spikes are detected. A deeper understanding of the processes that influence and generate spiking lead to better models, and hence more reliable interfaces.

Limits of Being Human

The most speculative piece of the emerging neurotechnology revolution is its effect on what it means to be a member of our species. If we could replicate the most cherished and specialized human brain functions, we could be able to augment our abilities. Imagine having four arms, six infrared detecting eyes, unlimited recall, or ultrasound perception. What if we embody all of our brain's abilities to perceive, reason, emote, or create into a desktop machine? The advanced technology needed to replicate the full abilities of a human brain in a box still seems very far off.

Such a "brain machine" would require explanations of the mechanisms through which the collective dynamics of networks link neurons to all behavior and technology to implement these mechanisms. Yet it is not unthinkably far. As more technology to replicate and extend emerges, and deeper understanding of the human brain emerges, closer approximations of machines to our own brain's most impressive capabilities will likely evolve, as they already have with smartphone applications that guide us to unfamiliar destinations, translate text from foreign tongues, or park our cars. These advances will precipitate a broad range of ethical challenges that require thoughtful debate and careful oversight to equalize opportunity and prevent abuse. Already there are present-day challenges as we navigate the use of pharmacological agents to augment our attention or change our mood, and ponder the use of limb prosthetics to enhance, for example, providing the ability to run faster or climb better than able-bodied people. As neurotechnology redefines what is possible, it will fundamentally reinvent the already profound debate about the boundary between man and machine.

Acknowledgments: I wish to thank my BrainGate colleagues who have enabled our contribution to the neurotechnology revolution. I also thank Miyoung Chun and the BAM group for inspirational discussions and the NIH B.R.A.I.N. committee members who, through their exceptional knowledge of neuroscience, assembled the networks in my brain that were necessary to write this chapter. Finally, I thank the VA, NIH, DARPA, TATRC, ONR, and NSF as well as the Samson and Israel Brain Foundations for their generous support of my research and Gary Marcus for his thoughtful editorial guidance.

Further Reading

Networks of Networks

Watts, D. J., and S. H. Strogatz. 1998. "Collective Dynamics of 'Small-World' Networks." *Nature* 393: 440–42. doi: 10.1038/30918.

Brains and Machines and the Future

The Economist. 2013. "Neuromorphic Computing: The Machine of a New Soul." *The Economist*, August 3. http://www.economist.com/news/science-and-technology/21582495 -computers-will-help-people-understand-brains-better-and-understanding-brains.

Brain-Computer Interfaces

Donoghue, J. P. 2008. "Bridging the Brain to the World: A Perspective on Neural Interface Systems." *Neuron* 60 (3): 511–21. doi: 10.1016/j.neuron.2008.10.037.

Hochberg, L. R., D. Bacher, B. Jarosiewicz, N. Y. Masse, J. D. Simeral, J. Vogel, S. Haddadin, J. Liu, S. S. Cash, P. van der Smagt, and J. P. Donoghue. 2012. "Reach and Grasp by People with Tetraplegia Using a Neurally Controlled Robotic Arm." *Nature* 485 (7398): 372–75.

Illes, Judy, and Barbara J. Sahakian, eds. 2011. *The Oxford Handbook of Neuroethics.* Oxford: Oxford University Press.

Wolpaw, Jonathan, and Elizabeth Winter Wolpaw, eds. 2011. *Brain-Computer Interfaces: Principles and Practice.* New York: Oxford University Press.

Optical Recording and Stimulation

Grienberger, Christine, and Arthur Konnerth. 2012. "Imaging Calcium in Neurons." *Neuron* 73 (5): 862–85.

Smedemark-Margulies, N., and J. G. Trapani. 2013. "Tools, Methods, and Applications for Optophysiology in Neuroscience." *Frontiers in Molecular Neuroscience* 6: 18.

THE MISWIRED BRAIN, GENES, AND MENTAL ILLNESS

Kevin J. Mitchell

If you go to the doctor with chronic abdominal distress, the physician and other technicians will perform a series of tests to try and figure out the cause. They may check inflammatory markers in your blood, test for allergies, perform a colonoscopy, take biopsies, test various enzyme levels, and so on. Ultimately, they may return a diagnosis of Crohn's disease or colon cancer or ulcerative colitis or any of a long list of discrete conditions that can manifest with similar symptoms. Knowing the cause will directly inform the treatment. In the event that a cause cannot be found, your illness will likely be labeled irritable bowel syndrome, which is simply a diagnosis of exclusion. It puts a name to your suffering, but offers no insight as to its cause.

In psychiatry, virtually all diagnoses are like that. Labels like major depressive disorder or schizophrenia or autistic spectrum disorder are defined by patterns of symptoms that often occur together, with a more or less typical course of illness. These are open constructs—defined not by a strict set of parameters, or the results of a particular test, but by reference to an exemplar. They give a name to the suffering of patients whose illness looks superficially similar based on the only data to which psychiatrists have access: patterns of behavior and patient reports of subjective states. They say nothing about causes because the field has known almost nothing about causes.

This is the main reason why almost no new drugs, with new mechanisms of action, have been developed for psychiatric conditions in over sixty years. This stands in stark contrast to the progress made in developing new drugs for heart disease, cancer, and other disorders. Such advances were made possible by increased knowledge of the underlying biology of these conditions, down to the molecular level. In psychiatry, efforts to elucidate the biological causes of various conditions have been frustrated not just by our lack of access to the affected tissue, but more fundamentally by the way in which such conditions are defined.

Psychiatrists recognize that the categories they have defined (and frequently redefine) do not represent "natural kinds"—groupings that really exist in nature as opposed to arbitrary human classifications. At best, they are useful terms to enable extrapolation of clinical experience between patients with similar symptoms. At worst, such similarities may be actively misleading, obscuring a diversity of underlying causes.

It has been hoped, particularly in the lead-up to the latest revisions of the international manuals of psychiatric diagnosis, that neuroscience could provide the insights required to recognize discrete diseases and to distinguish them by biological cause. These might include things like "excess serotonin in the nucleus accumbens" or "increased dopamine release in striatum" or "decreased functional connectivity between hippocampus and prefrontal cortex." So far, this approach has not been successful. There are no brain scan findings or biomarkers of any kind that can be used to assign a diagnosis of schizophrenia or bipolar disorder or autism.

A major reason for this failure is that the experiments designed to look for such distinguishing differences in brain properties rely on the diagnostic categories they hope to validate, or, indeed, replace. The best that neuroscientists can do is to compare the brains of groups of patients with "schizophrenia" or "autism" or other existing categorical labels to the brains of groups of controls. If these categories do not represent natural kinds, then lumping many cases together and looking for group differences that inform on primary causes is ultimately futile. Real differences underlying illness in particular subsets of patients will be swamped out if the causes of these conditions are actually heterogeneous.

Genetics

Where neuroscience has so far failed to distinguish psychiatric patients by cause, genetics is proving more incisive. From the time that conditions like schizophrenia and autism were first described it has been clear that they "run in families." Twin studies clearly show that this effect is largely due to shared genes, not shared environment. In neurodevelopmental disorders like autism and schizophrenia, genetic differences account for the vast majority of variance across the population in who develops these conditions.

The big news is that it is now finally possible to find those genes. Where previously we knew of rare mutations in a couple of genes that could lead to psychiatric disease, we now know of well over a hundred. Some of these mutations affect single genes, while others involve deletion or duplication of a section of a chromosome. The latter are known as copy number variants, or CNVs, because they change the number of copies of genes contained with the deleted or duplicated segment. Some such CNVs are associated with particular, rare conditions such as Angelman syndrome or Williams syndrome. But others can manifest with the symptoms associated with more "common" conditions like autism, schizophrenia, and epilepsy. While individually rare, this class of mutations may collectively account for 10 to 15 percent of cases of such disorders. Thanks to the development of new whole-genome sequencing technologies, it is now also much easier to detect a more subtle kind of mutation that only changes a single DNA nucleotide, affecting a single gene, to change the production or function of the encoded protein. Both CNVs and single-gene mutations can dramatically increase the risk of psychiatric disease, with anywhere from 10 percent to 100 percent of carriers affected by some psychiatric manifestation.

Several important and general insights have emerged from these studies, which are forcing a reconceptualization of psychiatric disorders. First, the mutations do not respect the arbitrary boundaries between current diagnostic categories. Particular mutations (as in the genes CNTNAP2, PCDH19, or SHANK3, for example) may manifest as schizophrenia in one person, as autism in another, and as intellectual disability or epilepsy in a third. This is not exceptional, this is the rule—there are no known mutations that manifest solely as a single psychiatric diagnostic category. This fits with recent epidemiological observations from very large-scale studies, which have shown a broad overlap in familial risk across psychiatric and neurological categories. The etiology of these supposedly distinct disorders is thus actually largely overlapping.

The second important point is that psychiatric conditions can arise due to a mutation in any one of a very large number of genes. Not ten or twenty, but likely on the order of at least a thousand. From an etiological perspective, psychiatric diagnostic categories are therefore not unitary conditions but umbrella terms describing possible outcomes of a very large number of distinct genetic syndromes. While these are individually rare, they are collectively numerous—common enough to account for the prevalence of conditions like schizophrenia and autism (each around 1 percent of the population, depending on definition).

Not all such mutations are inherited—at least not in the colloquial sense. Many arise de novo during the generation of egg or, more commonly, sperm cells. Thus even sporadic cases of disease, with no family history, can still have a genetic cause. High rates of de novo mutations and the large number of genes involved in these conditions explain why these disorders persist in the population, even though they increase mortality and reduce numbers of offspring, on average. Although many causal mutations are therefore not passed on to further generations, new ones emerge all the time to take their place.

It is important to emphasize the complex relationship between genotype and phenotype for most of these mutations. While many mutations dramatically increase risk of psychiatric disease, their effects can be quite variable, and most of them are also found, at lower frequency, in people who have never had occasion to seek any kind of psychiatric treatment. In many cases, genetic background will play an important part in determining the effects of a "primary" mutation. In others, multiple mutations may contribute to the emergence of disease. Nevertheless, for the growing subset of patients who carry a known pathogenic mutation, it is possible at least to assign it a major contributing role in their illness.

Genetic diagnoses of the cause of illness in any individual patient can have immediate and important psychological and social implications for the individual and his or her family. The diagnosis can also affect insurance coverage and can inform future reproductive decisions. In the longer term, the identification of pathogenic genetic mutations also opens a proven discovery route to elucidate the underlying biological mechanisms of disease and to distinguish patients on that basis.

From Genes to Biology

In the first instance, grouping patients by genetic lesion may reveal commonalities in symptom profiles or course of illness within small groups that were not apparent when such groups were hidden among the mass of patients. Defining the clinical sequelae of various genetic syndromes may thus help resolve some of the clinical heterogeneity associated with broad diagnostic categories and directly inform clinical management (figure 1).

The ability to segregate patients based on the biological cause of their illness also circumvents the previously intractable problem of heterogeneity for neuroscience research into psychiatric disease. This may allow researchers to define specific neurobiological defects associated with particular syndromes, where they have failed to do so for broad diagnostic categories. Neuroimaging of forty patients with 22q11 or 3q29 or 16p11.2 deletion syndromes may be far more informative than imaging four hundred patients with "schizophrenia." Highlighting particular neurochemical pathways or neural circuits affected in subgroups of patients may suggest avenues for treatment specific to those groups. These could involve the use of specific drugs, or, increasingly perhaps, direct intervention on the activity of neural circuits, as with deep brain stimulation for depression or obsessive-compulsive disorder, for example.

Ultimately, however, a full understanding of how a mutation results in psychiatric disease will require a far more detailed investigation, linking the levels of molecules, cells, circuits, networks, and brain systems. After all, how can changing one letter of the DNA code make someone paranoid, or manic, or suicidal? How can we bridge the gap between molecules and mind? Genetics provides a thread that can be followed across those levels, from cellular to animal models to humans.

One of the main findings from recent genetic discoveries is that many of the genes implicated in psychiatric illness normally function in processes of early brain development. Contrary to many artists' renditions, the cellular architecture of the brain is both incredibly complicated and exquisitely organized. There are many hundreds of distinct types of nerve cells, in thousands of different areas, all of which have to be distributed and connected to each other in highly specific ways. The fact

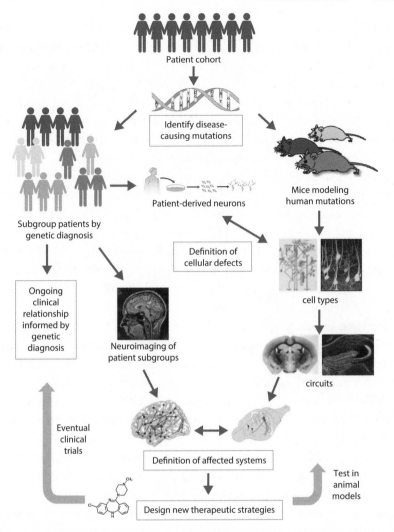

Labels within figure: Patient cohort; Identify disease-causing mutations; Patient-derived neurons; Mice modeling human mutations; Subgroup patients by genetic diagnosis; Definition of cellular defects; cell types; Ongoing clinical relationship informed by genetic diagnosis; Neuroimaging of patient subgroups; circuits; Eventual clinical trials; Definition of affected systems; Test in animal models; Design new therapeutic strategies

Figure 1. A genotype-first approach to psychiatric disorders. Genetic analyses can be used to discern subgroups of patients within broad diagnostic categories. Identification of specific mutations allows the generation of cellular and animal models of direct etiological validity, which can be analyzed to elucidate the cascading effects of mutations, from molecular and cellular levels to the ultimate effects on neural circuits and systems. Combined with neuroimaging and other analyses of genetically defined subgroups of patients, these investigations may reveal the nature of the emergent pathophysiological states and suggest possible therapeutic strategies, which can be tested in animal models and, eventually, in clinical trials in defined cohorts. Stem cell panel courtesy of Jamie Simon and Fred H. Gage, Ph.D., Salk Institute for Biological Studies. Human brain network panel courtesy of NITRC; BrainNet Viewer. Cell types panel reproduced from Mitchell et al. *BMC Biology* 2011, 9:76.

that this circuitry self-assembles, based on a developmental program encoded in the genome, is one of the most remarkable feats of evolution.

But that program is vulnerable. It involves the products of many thousands of genes—proteins that specify where cells will migrate, where their nerve fibers project, which cells they will connect with, and how those connections will change with use. It is mutations in those kinds of genes that are turning up in many patients with psychiatric illness.

Animal models are proving invaluable in working out how mutations affecting neural development or brain plasticity can ultimately result in the emergence of pathological brain states. A recurring theme is the implication of an imbalance in the functions of two main classes of neurons: those whose signals excite other neurons, encouraging them to fire an electrical signal of their own, and those that inhibit other neurons, dampening down their electrical activity. This imbalance manifests most obviously as epilepsy—the uncontrolled firing of large populations of neurons. But inhibitory neurons do far more than simply prevent runaway excitation. They also, crucially, control many aspects of information processing in neural circuits, such as filtering, gain control, and temporal and spatial integration. In addition, they orchestrate the synchronous and oscillatory firing of ensembles of excitatory neurons, which in turn is a central mechanism mediating communication between brain areas. The activity of entire brain systems thus derives from the emergent properties of synapses, cells, and microcircuits.

Studies in mice carrying mutations associated with psychiatric illness are now shedding light on the diverse primary defects that arise due to mutations in various genes and also highlighting the cascading effects that emerge over subsequent neural development, ultimately pushing the brain into a pathological state.

Consider Fragile X syndrome, which is caused by mutation in a single gene, on the X chromosome, and is one of the most common causes of intellectual disability. Many Fragile X patients also present with the symptoms of autism, and Fragile X mutations account for 3 to 4 percent of autism cases. This syndrome was described in 1943, but the mutated gene was not molecularly identified until 1991. Based on findings in cellular and animal models, interpreted in the context of a vast literature from basic neuroscience studies, the function of the mutated protein (called FMRP) was elucidated. It works at neuronal synapses to put the brakes on a biochemical process that generates new proteins and mediates synaptic plasticity—changing the strengths of connections between neurons.

When the gene encoding FMRP is mutated in mice, the process of synaptic plasticity is misregulated, excitatory neurons become hyperconnected, at the cellular level, and the cortex becomes hyperexcitable, with altered patterns of rhythmic activity. These changes are associated with cognitive defects, impaired social interaction, hyperactivity, auditory hypersensitivity, and audiogenic seizures, mimicking many aspects of Fragile X syndrome in humans.

The detailed knowledge of how FMRP functions suggested a therapeutic approach: if the brakes on the process of synaptic plasticity were not working so well, then perhaps this pathway could be rebalanced by taking our foot off the gas. FMRP normally antagonizes a pathway that is activated by a protein that senses the level of activity between neurons—a metabotropic glutamate receptor. In the absence of FMRP, that pathway is overactive. Lowering the amount of the metabotropic glutamate receptor, or blocking its function with drugs, proved remarkably successful in reversing many of the effects of mutation of the Fragile X gene in mice, from the cellular level to the physiological and behavioral levels. Drugs that block this metabotropic glutamate receptor are now in clinical trials for Fragile X syndrome.

Although still at an early stage, these efforts illustrate the core concept in medicine of developing therapeutics based on detailed biological knowledge, as opposed to serendipity or random screening of chemical compounds. For a field where no drugs with new mechanisms of action have been developed for over sixty years, the seismic nature of this paradigm shift cannot be overstated.

This example also highlights the importance of *individual* genetic diagnoses and personalized treatment. Even if such drugs can prevent or reverse some of the symptoms in Fragile X patients, they may not be effective in other cases of intellectual disability or autism. They may, in fact, be contraindicated for some patients. Tuberous sclerosis is another genetic syndrome often characterized by symptoms of autism. Mutation of the responsible gene also affects the molecular pathways of synaptic plasticity, but in this case the biochemical effects are directly opposite to those observed in Fragile X syndrome. Although dysfunction of the biochemical pathway in either direction can result in autism, it is clearly important to know which is involved in any given patient, as drugs that ameliorate Fragile X syndrome may exacerbate symptoms in a patient with a tuberous sclerosis mutation. A similar situation holds for some specific genetic causes of epilepsy, such as Dravet syndrome, where certain anticonvulsants are contraindicated due to interactions with the sodium channel, which is mutated in ~80 percent of patients with this condition. Knowing the cause informs the treatment.

We are on the cusp of a real revolution in the treatment of mental illness. Genetics will transform psychiatry from a discipline based on arbitrary diagnostic constructs capturing only similarities in superficial symptoms to one that distinguishes many patients based on the root causes of their illness. It also affords a proven route to discover the underlying biology of brain-based disorders and ultimately to provide a genuinely personalized approach to treatment for psychiatric patients.

Further Reading

Arguello, P. A., and J. A. Gogos. 2012. "Genetic and Cognitive Windows into Circuit Mechanisms of Psychiatric Disease." *Trends in Neurosciences* 35: 3–13.

Greenberg, G. 2013. *The Book of Woe: The DSM and the Unmaking of Psychiatry*. New York: Blue Rider Press.

Keller, M. C., and G. Miller. 2006. "Resolving the Paradox of Common, Harmful, Heritable Mental Disorders: Which Evolutionary Genetic Models Work Best?" *Behavioral and Brain Sciences* 29 (4): 385–404, discussion at 405–52.

Krueger, D. D., and M. F. Bear. 2011. "Toward Fulfilling the Promise of Molecular Medicine in Fragile X Syndrome." *Annual Review of Medicine* 62: 411–29.

Mitchell, K. J. 2011. "The Genetics of Neurodevelopmental Disease." *Current Opinion in Neurobiology* 21 (1): 197–203.

NEURAL DUST

AN UNTETHERED APPROACH TO CHRONIC
BRAIN-MACHINE INTERFACES

Michel M. Maharbiz

With Dongjin Seo, Jose M. Carmena, Jan M. Rabaey, and Elad Alon

Brain-machine interface (BMI) technology (see chapter by Donoghue) aims to improve the quality of life for those suffering from paralysis and neurological conditions such as amyotrophic lateral sclerosis and stroke. Half a century of scientific and engineering effort has yielded a vast body of knowledge and a closely related set of tools for interfacing the mammalian brain that should lead to clinically viable applications. Yet two main challenges remain: (1) engineering fully implantable, untethered, clinically viable neural interfaces that last a lifetime, and (2) boosting performance to achieve skillful control and dexterity of the prosthetic device to a level that will justify the risk:benefit ratio of having such a device implanted.

Creating lasting, durable, untethered interfaces raises a variety of issues, ranging from the nature of the physical substrate (avoiding the biotic and abiotic effects that presumably lead to performance degradation at the electrode-tissue interface, the density and spatial coverage of the sensing sites), the type of signals measured, and the computation and communication capabilities (how much signal processing on-chip data to transmit wirelessly) under the power budget of the whole system.

The second challenge gravitates toward the question of what level of control and dexterity of the prosthetic device can be achieved with the signals provided by the neural interface that will justify implanting this device in the brain? One important part that has not received much attention until recently is the encoding of sensory feedback from the prosthetic device by directly stimulating sensory areas in the brain. The idea is to complement visual feedback so that the user can also *feel* the environment. This has been supported by recent examples using electrical microstimulation during active sensing tasks. Another part that is going to play a pivotal role in future BMI developments is the view of the BMI as a system in which *both* the neurons and the algorithms decoding neural signals adapt together toward accelerating learning, improving system performance, and providing the BMI with natural motor memory-like properties. Ultimately, the goal is to achieve a quantum-leap increase in the controllability of neuroprosthetic devices that should allow a patient to perform tasks of daily living in a natural and effortless way.

Addressing these important challenges is critical for BMIs to have a broad and important clinical impact. In this chapter we focus on the first challenge, specifically on introducing a new technology that will radically increase the number of recording sites from the brain while eliminating transcranial wires and enabling lifetime-scale operation.

A typical intracortical BMI system is comprised of four different subsystems acting together—namely, the neural interface that measures the extracellular activity from populations of neurons (action potentials and local field potentials [LFP]) in cortical areas of the brain, the decoding algorithm that translates these signals into motor commands, the prosthetic device that executes these motor signals, and the feedback about the state of the prosthetic device.

Currently, the majority of neural recording is done through the direct measurement of electrical potential changes near neurons during depolarization events called *action potentials*. Few other approaches are sufficiently localized (high spatial resolution) and fast enough (high temporal resolution) to be able to capture action potentials single neurons produce. Optogenetic methods are an exciting alternative, but they have not yet been modified successfully for widespread *clinical* use as they nominally involve genetic manipulation of the host cells. There are also numerous clinically useful modalities with which one can extract information from the brain. Advances in imaging technologies such as functional magnetic resonance imaging (fMRI), electroencephalograph (EEG), positron emission tomography (PET), and magnetoencephalograph (MEG), have provided a wealth of information about collective behaviors of groups of neurons. Numerous efforts are focusing on intra- and extracellular electrophysiological recording stimulation, optical recording, optogenetic stimulation, optoelectrical, and electroacoustic methods to perturb and record the individual activity of neurons in large (and, hopefully scalable) ensembles. All recording technologies embody some fundamental trade-off between temporal or spatial resolution, portability, power requirements, invasiveness, and so on.

While the specifics vary across several prominent technologies, all extracellular electrical recording interfaces share several characteristics:

- a physical, electrical connection between the active area inside the brain and electronic circuits outside the skull
- a practical upper bound of several hundred implantable recording sites
- the development of a biological response around the implanted electrodes that degrades recording performance over time. To date, chronic clinical neural implants have proved to be successful in the short range (months to a few years, but not longer) and only for a small number (~10s) of channels.

Is there a way to embed very tiny recording devices in the brain such that we could radically increase the number of recording sites while eliminating transcranial wires and enabling lifetime-scale operation? We believe the answer is yes. In what follows, we sketch out the technical rationale for why this may be possible, with the caveat that this work is in its infancy.

Introductory Concepts

The technology we propose leverages the amazing advances in silicon electronics made possible by the information revolution, and the set of related manufacturing processes for building silicon chips that contain hundreds of millions to billions of nanoscale switches known as *transistors*. Such transistors, operating in concert as *circuits*, allow chips to measure the world around them (e.g., how fast you shake your phone), communicate wirelessly (wifi networks and mobile phone calls), produce video, and uncounted other communication, computation, and sensing functions. The technical name often used for this technology is Complementary Metal Oxide Semiconductor technology, or CMOS for short.

A second technology leveraged below is that of *piezoelectricity*. Piezoelectricity has a long and distinguished history and technical literature. In short, certain crystals, when stretched, will produce an electrical voltage. These same crystals will also compress when a voltage is applied to the material. The same materials are also mechanically very low loss; that is, if given a mechanical compression or stretched and released, they will vibrate (just like a tuning fork once struck) for quite a while at *ultrasonic frequencies* before the energy is lost as heat (and the vibration stops). As a comparison, humans can hear mechanical vibrations with frequencies up to about 15 kHz (the vibration wiggles 15,000 times per second); piezo crystals can vibrate well into the MHz range (millions of times per second). These observations led to the use of tiny piezoelectric crystals as high-frequency timers, ultrasound microphones (for medical imaging), ultrasonic "speakers," and many other applications.

The Neural Dust Paradigm

In its simplest form, *neural dust* consists of a piezoelectric crystal coupled with a very small CMOS recording chip (figure 1). The crystal, or *transducer*, is used as an energy harvesting unit; ultrasonic energy impinging on the crystal causes it to vibrate, producing a voltage that can supply electrical power to the CMOS chip. Such crystals, at larger sizes, are a mainstay of modern electronics. Mounted on the crystal is an infinitesimally small CMOS chip with surface electrodes for neural signal acquisition. The chip uses the crystal to report recorded information back to a distant interrogator by reflecting and modulating either amplitude, frequency, or phase of the impinging ultrasound wave. We will often refer to single neural dust systems as *nodes*. The modulation mechanism of each node is detailed in the later sections. There exist fundamental system design trade-offs and ultimate size, power, and bandwidth scaling limits of systems built from low-power CMOS coupled with ultrasonic power delivery and backscatter communication.

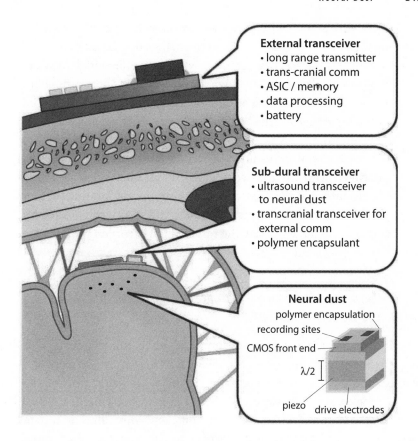

Figure 1. Neural dust system diagram showing the placement of ultrasonic interrogator under the skull and the independent neural dust-sensing node dispersed throughout the brain.

Mechanisms for Powering and Communicating with Implants

The requirements for any computational platform interfacing with microelectrodes in order to acquire neural signals of use in high-quality motor control are fairly stringent. The two primary constraints on the implanted device are device size and power. These are discussed in greater detail below, but briefly:

- implants placed into cortical tissue with scales larger than one or two cell diameters have well-documented tissue responses that are ultimately detrimental to performance and occur on the timescale of months; some debate exists as to what role mechanical anchoring outside the cortex plays in performance degradation

- potentials (extracellular or otherwise) are differential measurements, so as devices scale down and the distance between recording points decreases, the absolute magnitude of the measured potential increases, which requires scaling down front-end noise; this requires power (that is, increasing power lowers the noise floor for a fixed bandwidth). Additionally, to eliminate the risk of infection associated with the transcutaneous/transcranial wires for broadcasting information from and powering the device, such tethers should be avoided as much as possible; a global wireless hub is therefore essential to relay the information recorded by the device through the skull.

For Very Small Implants, Electromagnetic Waves Are Not a Good Choice to Couple Signals in to and out of the Brain

The most popular existing wireless transcutaneous energy transfer technique relies on electromagnetic (EM) fields or waves transferring information and power. Energy coupling via magnetic fields, for example, has been used in a wide variety of medical applications and is the principal source of power for cochlear implants. As EM requires no moving parts or the need for chemical processing or temperature gradients, it is considered more robust and stable than other forms energy scavenging. When used in-body, however, the total power that can be "beamed in" is restricted by the potential adverse health effects associated with heating of tissue (as EM fields cross tissue, some heating occurs). This is regulated by the FCC, and IEEE-recommended levels are well known (roughly, you must send in less power than that required to heat tissue by 1° C.)

Consider, in this context, the problem of transmitting EM power to (and information from) very small circuits embedded in tissue. There are two problems. The first arises from the speed of light itself. Because EM waves are so fast (~300,000,000 m/s), any structures in the 1 μm to 1 mm size range would be resonant only at very high frequencies (> 10 GHz). At these frequencies, the loss of the EM signal is very high. In addition, the EM fields lose quite a bit of power when traveling through tissue (this, in fact, is what leads to the heating of the tissue).

EM waves are not the only way to transfer energy, however. Among other ways are ultrasonic waves: sound (pressure) waves that oscillate at frequencies above human hearing. The speed of ultrasound (~1,500 m/s) in the brain is much, much lower than that of EM waves (speed of light given earlier). Moreover, the ultrasound loss in the tissue is also significantly reduced compared to EM. Thus, lowered tissue loss coupled with the slower speed of sound means that a piezocrystal of a given size will "couple in" more power from ultrasonic waves imping-ing on it than an EM coil or antenna of a similar size (at these small scales). In short, our calculations show that similarly sized devices would "capture" ~10 million times more energy using ultrasonic waves instead of EM!

Coupling power in is not the only problem. Can we build tiny electri-cal recorders of neural signals at this scale? A typical extracellular elec-trophysiological recording of neural activity in tissue records electrical potential differences between one electrode placed in-tissue near the neural activity and a second electrode "far away." This is not the case for our motes: both electrodes are on board the tiny device and are placed very close together. This makes it very hard to measure the tiny electri-cal changes that arise across these electrodes. To some extent, the tiny electronics can be made more sensitive by pumping in more power. This creates a race to the bottom: smaller motes capture less power but need more power to record the tiny signals. Somewhere around a 50 μm diameter, our calculations show you cannot deliver enough power to power the sensor electronics.

The second challenge involves simultaneous gathering and distin-guishing information from multiple sensing sites. For functional neural mapping applications, which will likely require the full, digitized neu-ral signal, each node will generate > 1 kbps of neural data that needs to be continuously streamed to the interrogator. This will likely mean using multiple interrogators and possibly operating different dust nodes at different transmit frequencies. In contrast, for BMI applications, we only need to be able to resolve the occurrence of a neural spike, which significantly reduces the burden on data postprocessing.

What Would the Recording Electronics on the Node Look Like?

In addition to the challenge of delivering power, managing noise limits, and communicating back information, there is the challenge of designing the "optimum" circuit within the dust mote that performs the electrical recording. A quick glance at the problem highlights the challenge. First, there isn't a lot of room on each node to design a "complete" amplification and digitization front end with today's (or near future) circuit technology; such a circuit would involve hundreds to thousands of transistors and would simply not fit. Second, there is likely not enough power to drive a complex circuit of this type. Our solution to this is to build an incredibly simple circuit: one transistor! Unlike existing systems that attempt to amplify, digitize, and communicate the information in a sophisticated way, our approach relies on extracellular potential differences arising across two electrodes (which are connected to the transistor's terminals) being used to "gate" the current flowing across the single transistor. Changes in this current, in turn, affect how the piezocrystal "rings" (in a very small way); these changes then affect the signal that bounces back from the piezocrystal to the transmitter. As an imperfect analogy, picture a ringing tuning fork (and a friend hearing the sound); in response to signals only you can hear, your hand lightly touches the tines, changing what your friend hears; in this example, you are the transistor, the tines the piezocrystal, and your friend the distant receiver.

Conclusions

In short, it appears feasible to build ultrasmall (~50 μm), untethered, neural recording devices powered by ultrasound. Three challenges must be overcome. The first is the design and demonstration of CMOS circuitry suitable for operating within the extreme constraints of decreasing available power and decreasing signals with scale. The second challenge is the integration of extremely small piezoelectric transducers and CMOS electronics in a properly encapsulated package. The above discussion assumed the entire neural dust implant was encapsulated in an inert polymer or insulator film (a variety of such coatings are used routinely in neural recording devices; these include parylene, polyimide, silicon nitride, and silicon dioxide, among others) while exposing two recording electrodes to the brain. The third challenge arises in the design and implementation of suitably sensitive subcranial transceivers that can operate at low power (to avoid heating between skull and brain).

In addition to these three challenges, there is the additional problem of *how* to deliver neural dust nodes into the cortex. The most direct approach would be to implant them at the tips of fine-wire arrays similar to those already used for neural recording. Neural dust nodes would be fabricated or postfab assembled on the tips of array shanks, held there by surface tension or resorbable layers. Once inserted and free, the array shanks would be withdrawn, allowing the tissue to heal. Kinetic delivery might also be an option, but there is no existing data to evaluate what effect such a method would have on brain tissue or the devices themselves. All of these remain open challenges. If they can be met, as we suggest, these devices would present a completely new route to stable, long-term brain recording, something of immense importance to both neuroscience and the development of clinically relevant neuroprosthetics.

Further Reading

Biederman, W., D. J. Yeager, N. Narevsky, A. C. Koralek, J. M. Carmena, E. Alon, and J. M. Rabaey. 2013. "A Fully-Integrated, Miniaturized (0.125 mm²) 10.5 μW Wireless Neural Sensor." *IEEE Journal of Solid-State Circuits* 48 (4): 960, 970. doi: 10.1109/JSSC.2013.2238994.

Carmena, J. M. 2012. "How to Control a Prosthesis with Your Mind." *IEEE Spectrum* 27 (February).

Green, Andrea M., and John F. Kalaska. 2011. "Learning to Move Machines with the Mind." *Trends in Neuroscience* 34 (2): 61–75. doi: 10.1016/j.tins.2010.11.003.

Seo, Dongjin, Jose M. Carmena, Jan M. Rabaey, Elad Alon, and Michel M. Maharbiz. 2013. "Neural Dust: An Ultrasonic, Low Power Solution for Chronic Brain-Machine Interfaces." *arXiv*: 1307.2196.

AFTERWORD

NEUROSCIENCE IN 2064

A LOOK AT THE LAST CENTURY

As told to

Christof Koch and Gary Marcus

On a warm summer evening, not long ago, while we, the authors, were engaged in a spirited debate about the nature of consciousness, a traveler, going by the name of Lem, appeared, claiming to be from the future; at first, we were skeptical. But his recollections were vivid, and detailed, and more than that, internally consistent. Try as we could, we couldn't break his story; he claimed to be from the year 2064, and his knowledge of neuroscience seemed to be exceptional. Over time, we began to believe that his reports were authentic; in what is below, we have transcribed his story as near as we can recall it.

One hundred years ago, in 1964, the United States and the Soviet Union were jockeying for world supremacy, "computers" still meant human beings, trained to carry out long chains of calculations, and gas guzzlers dominated the highways. Global warming and nanotechnology were not even in the vocabulary, and a British band known as The Beatles had just arrived in America.

What difference one hundred years make! Extreme weather and greatly diminished fossil fuels, the decline of the American and Russian empires, the rise of the Chinese Dragon, and the widespread intrusion of artificial intelligence agents into daily life has transformed the stable, dichotomous Cold War world of 1964 into a more splintered world, vibrant yet at the edge of chaos in its own way. Some of us live longer and more healthfully than our ancestors, as dozens of once-deadly diseases have been cured. Yet the bulk of mankind still lives less than four score and ten years; and the promises of trans-humanists to extend the maximal life span past 120 years have thus far proved illusory.

Molecular biology has finally delivered on the early promises of the Human Genome Project, albeit decades later than forecast. Previous monolithic diseases, such as breast cancer, brain cancer, depression, dementia and autism, have splintered off into a myriad of more specific pathologies, defined not so much by common behavioral phenotypes but by shared mutations, molecular pathways and biochemical mechanisms. In combination with cheap, reliable, and fast genetic tests, the age of personalized medicine, long trumpeted by Leroy Hood, Craig Venter, and other pioneers, arrived in which familial predispositions to behavioral traits, pharmacological interventions, and diseases, permit much more targeted interventions.

Bioterrorism has occasionally struck, but the combination of personal genomics, personal immunizers, and a ubiquitous surveillance state has largely kept the population safe.

Advances in the brain sciences have been in many ways even more impressive; a hundred years ago, humanity knew that the brain—and not the heart or liver—was the seat of the mind, but little about how neural tissue governed perception, comprehension, or consciousness; brain-machine interfaces, now common, did not even figure in the most popular science fiction television program of the day (*Star Trek*). If our understanding of neuroscience is still incomplete, it is shocking how much progress there has been. Yet one also forgets that the seeds for our modern understanding were already in place.

The Romantic Era of Neuroscience: 1964

The first blossoming of the romantic era in neuroscience started almost two centuries ago. It was powered by two technologies, the *optical microscope* and the refinement of *chemical dyes*, in particular Golgi's staining method of using silver chromate salt. Together, these allowed Santiago Ramón y Cajal to visualize in stunning detail the circuitry of the nervous systems in animals and people, demonstrating in aesthetically pleasing images that brains, like kidneys, hearts and all other biological organs, are composed of a myriad of discrete, cellular units, neurons, and their supporting actors, glial and astrocytes. Neurons, he discovered, came in a dizzying variety of shapes, sizes and geometries.

Later, the *electron microscope* established beyond doubt that nerve cells were linked at discrete specialized junctions, chemical and electrical synapses, and the *microelectrode* recorded the electrical activity of individual nerve cells. In 1963, the Nobel Prize was awarded to John Eccles for discovering the discrete (quantal) nature of synaptic transmission, and to Alan Hodgkin and Andrew Huxley for describing the sodium and potassium membrane currents that power the electrical impulse, the famed action potential or spike, as it travels along the axon. The mathematical formalism they pioneered has proved enduring; the reign of the Hodgkin-Huxley equations describing the biophysics of individual nerve cells would last until they were replaced by molecular dynamics model in the 2020s.

The next major advance came from electrical recordings from anesthetized and, subsequently, from awake and behaving animals with microelectrodes coupled to miniaturized differential amplifiers (and loudspeakers), which made the hitherto silent brain come alive with the staccato sounds of spiking nerve cells. In their classical 1959 and 1962 studies, David Hubel and Torsten Wiesel discovered the selectivity of visual cortical cells to the orientation of lines that the animal looked at. This work in turn launched the bold exploration of the higher order visual cortex that culminated in the late 1960s with the discovery of individual neurons that responded preferentially to faces.

Clinical studies, always a fecund source of knowledge about human nature, had given birth to neurology and to neurosurgery, both of which contributed to neuroscience. The neurologist Paul Broca had first inferred in 1861 from a singular patient that a specific region of the left inferior frontal gyrus is critical to speech. By the 1930s and 1940s, the neurosurgeon Wilder Penfield had stimulated the exposed cortex of epileptic patients with electrodes, thereby triggering simple visual percepts, movements, or vividly recalled memories, again and again. This was a compelling demonstration of the intimate link between the physical brain and the subjective mind.

In mathematical logic, Warren McCulloch and Walter Pitts demonstrated back in 1943 that interconnected networks of very simple neuron-like units could compute any logical expression. In conjunction with the Church-Turing thesis formalizing what is algorithmically computable, theoreticians and engineers established a foothold into the all-important challenge of conceptualizing how the brain could think, reason, and remember. Whereas René Descartes, three hundred years earlier, needed to postulate a vague cognitive substance (*res cogitans*) that did the thinking for people (famously, not for animals), computer scientists such as Frank Rosenblatt, inspired by McCulloch and Pitts, began taking the first tentative steps toward building computer simulations of brain-like circuits. If "Perceptrons," the single-layered neural networks of the 1960s, seem comically simplistic in hindsight, it must be remembered that such simple networks ultimately inspired a revolution. This period of boundless optimism and excitement was cross-fertilized by the launch of Artificial Intelligence in 1955 at Dartmouth College.

Neurophysiologists, computer scientists, and psychologists alike naively imagined that an understanding of the brain was near to hand. Of course, we now know that robust artificial intelligence took a century, not a few decades, to come about, and that neither psychology nor neuroscience was close to having reached the maturity that physics has. But the roots were all there. Nobody really knew remotely how the human brain worked, or how to emulate it, yet the revolution was well underway.

Neuroscience Becomes Big Science: 2014

Fifty years on, studying the brain was no longer a niche field but a full-on movement. The US-based Society for Neuroscience alone had more than forty thousand members, annual funding was well in excess of several billion dollars, and writers, journalists, and an inchoate neuro-industry all thrived on the public interest in the brain.

One major advance was molecular. Scientists had discerned the structure and function of ionic channels and receptors, the miniaturized stochastic switches and modulators embedded in the bilipid membrane that endows neurons with their ability to process information, to shape and guide action potentials along axons, and to release neurotransmitters. Also well understood was the action of sensory receptors that transduce the signals impinging onto the body—photons of light, sound perturbations in the air, or molecules of some odorant—into electrical activity. Indeed, neuroscientists had tracked down how single nucleotide changes in the DNA that encodes one or another photo-pigment protein in the retina impacted the way a subject perceives color.

The molecular revolutions of the day are perhaps best exemplified by the Nobel Prize–winning work of Eric Kandel, which elucidated how the sea slug *Aplysia* learns the gill-withdrawal reflex, the first form of long-term memory to be well understood. It demonstrated the importance of protein synthesis and changes in synaptic connectivity in long-term memory. Kandel's work furthered the growing realization that much of memory is encoded in the specific pattern and strength of connectivity among large ensembles of active neurons (as hypothesized already in 1895 by none other than Sigmund Freud), though the many ways in which memories could be stored *within* an individual neuron were not yet recognized. As Kandel and his contemporaries began to realize, the rules that determine how the influence that one synapse brings to bear on the neuron it is connected to, its *weight*, is up- or downward adjusted depends on the relative timing of the arrival of the pre- and postsynaptic electrical activity. (Cleverly, this gives individual synapses a rudimentary capacity for learning causal relationships, in which event **A** is followed by event **B** but never the other way around.) In 2013, the group of another Nobel laureate, Susumu Tonegawa, became the first to induce a false memory into mice by directly manipulating the underlying neural engram in their hippocampus. A great many molecular details—of the underlying neurotransmitters, second messenger systems, protein kinases, ionic channels, and transcription factors—were all steadily being filled in, even though the overall logic of the brain remained a mystery.

Two techniques proved transformative. First, in the 1980s, the physics of nuclear magnetic resonance was exploited to routinely, reliably, and safely image the static, anatomical structure of the human body by bombarding subjects with radio waves while they were lying inside powerful magnets. Applied to the brain, magnetic resonance imaging (MRI) revolutionized neurology. In the 1990s, MRI was refined to image the *functional* architecture of the active brain with spatiotemporal resolution at the scale of millimeters and seconds. Although the popular images of that time seem laughably crude by contemporary standards, they gave birth to the field of cognitive neuroscience as scientists began to investigate the neural basis of seeing, hearing, feeling, thinking, and remembering. Wars broke out about the "localization" hypothesis when many neuroscientists rejuvenated the old phrenologist program of linking specific mental faculties to specific parts of the brain, identifying more than one hundred brain regions on the basis of functional specializations. By 2014, theories of cognitive neuroscience began to grow in sophistication, as investigators realized that these specific regions formed parts of larger, more complex networks, which at that time eluded understanding. Only a few brain scientists were concerned with the coupling between fMRI signals, reflecting the power consumption of the brain at a sedate pace of seconds, and the switching in the underlying neural lattice at the millisecond scale. Indeed, the elementary spatial unit of brain imaging, *voxels*, at that time about 2 x 2 x 2 mm^3, encompasses about one million highly diverse neurons, glial cells, and astrocytes and ten billion synapses, firing two to twenty times within one MRI scan cycle, way too coarse to infer neuronal mechanism, akin to trying understanding language by listening to a smeared-out recording of the chattering among all the spectators at a sports arena. And few people had any conception of how important glial cells would turn out to be. Techniques like EEG and MEG were better temporally; they recorded electrical and magnetic fields with millisecond precision, but with even less spatial precision. The blurriness of these instruments was mirrored by the primitive and edentate tools used to safely perturb the human brain—electrical stimulation in patients, and extracranial electromagnetic fields and drugs in volunteers.

The other major advance fifty years ago was the birth of opto- and pharmaco-genetics, methods that delicately, transiently, reversibly, and invasively control defined events in defined cell types at defined times, initially in a few model organisms—the worm, the fly, and the mouse. Equipped with these tools for perturbing the brain, scientists systematically moved from correlation to causation, from observing that this circuit is activated whenever the subject is contemplating a decision to inferring that this circuit is necessary for decision making or that those neurons mark a particular memory. By the early 2020s, the complete logic of thalamo-cortical circuits could be manipulated, in hindsight a tipping point in our ability to bridge the gap between cortex and theories of its universal and particular functions.

An enormous amount of work characterized how sensory systems process their information and represent it in the cortical tissue. Silicon microelectrodes and live brain imaging using fluorescent dyes and genetically encoded proxy markers of electrical activity allowed intrepid neuroscientists to track the electrical activity of hundreds of neurons in the behaving animal simultaneously, a significant increase over the previous decades in which the brain was sampled by a single wire. Theoreticians could thereby infer from the firing of neurons the probabilistic manner in which the nervous system represents the visual, auditory, and olfactory environment, as well the animal's physical location, the animal's uncertainty in the face of a perceptual or a subjective decision, and even the presence of familiar individuals such as celebrities.

Yet despite these advances combined with the exponential increase in relevant data and the efforts of the brightest minds on the planet, comprehension of the brain's circuits in health and disease increased sublinearly. Even the smallest of all multicellular "model organisms," the roundworm *C. elegans*, whose nervous system contains a mere 302 neurons, was scarcely understood as a whole. Hundreds of worm specialists focused on isolated reductionist accounts of one function or another. Yet no one attempted to integrate all this knowledge into a single, coherent, comprehensive, holistic, and explanatory framework. Nor had any brain disease yet been cured. Many in the rapidly growing elderly population faced symptoms of dementia, yet little could be done to slow down the ravages of the disease; it must have been heartbreaking to witness. When the once dominant *Diagnostic and Statistical Manual of Mental Disorders*—at the time the psychiatrist's bible for treating patients with mental afflictions—appeared in its fifth edition in 2013, it did not list a *single* biomarker nor a single fMRI diagnostic criterion. If you were depressed, heard voices, or felt persecuted in the early twenty-first century, your only options were to talk to a therapist, fill out questionnaires, and take little-understood drugs that swamped your brain and had untold side effects.

In fairness, such slow progress was inevitable. Historically, science had been most successful when studying isolated systems with reduced degrees of freedom that tamed their complexity: a marble rolling down an inclined plane, a planet that plows its orbit around its center star, a lone electron in a magnetic field, a double strand of DNA. Even though it was obvious that living systems were characterized by large numbers of highly heterogeneous components, be they proteins, genes, or nerve cells, it was far from obvious how to deal with that complexity. A fundamental problem in the brain sciences has always been the numerous ways in which components interact causally across a large spectrum of space-time, from nanometers to meters and from microseconds to years. A complete understanding demands that a large fraction of these interactions be experimentally or computationally probed. This is fiendishly difficult. Bioinformaticians had few clues about how to integrate computations that spanned so many scales of time and space, and they lacked the relevant hardware, as cloud computers were primitive.

It was already becoming clear just how hard the problem was; even today no single human understands how the brain works at anything but an abstract and highly simplified level. Nature provides few shortcuts; a complete understanding of the brain comes not from any one experiment but from the integration of thousands of experiments that bridge many levels. Engineered systems such as spacecraft or computers that contain billions (then) or trillions (now) of discrete components are quite different. They are purposefully built to *limit* the interactions among the parts to a small number. Thus design rules for the layout of integrated electronic circuits impose a minimum distance between wires and other components to eliminate coupling, and the power supply is kept separate from computing, with computing separate from memory. Yet nervous systems interdigitate practically everything, from power supply to computation to memory. Nature couldn't have made herself more difficult to understand if she had tried. Early twenty-first-century scientists had begun to recognize this complexity but were unprepared and unable to deal with its consequences.

The next major revolution was not technological, but organizational. A private American initiative, the Allen Institute for Brain Science, taking cues from the biotechnology industry, was the first to approach neuroscience as "Big Science," moving from a model oriented around autonomous "star" investigators toward a team-based approach in which several hundred scientists from molecular biology, anatomy, physiology, genomics, optics, physics, and informatics worked together on industrial-scale projects, the first several of which had been launched by 2014 (see the chapter by Koch and colleagues, this volume). One generated the complete ontology of cortical cell types—the shape of their dendritic tree, the near- and far-flung target zones of their axons, the genes they express, their electrical behavior, and the rules governing their connectivities in the mouse and the human brain. The other was the construction of brain observatories—cerebroscopes—to record, make publicly accessible, analyze, and model the cellular events in the cortico-thalamic system underlying visual information processing in behaving mice. Other, even larger enterprises were spawned in the 2020s, as China and India became scientific world powers.

Also notable from that time was the publicly funded European Human Brain Project, which built a series of ever-larger supercomputer facilities to simulate, at the cellular, and, ultimately, at the subcellular level, the biophysics of neurons and their supporting cellular actors, in brains of increasing size, from the mouse to the human brain. Early on, their combination of morphological, anatomical, and physiological knowledge yielded an electrical model of a cortical column in rodents, a proof-of-principle that the electrodynamics of a chunk of brain matter could be understood by combing detailed biological knowledge with sufficient computational resources. The vision of a gigantic computer model of the human brain with the promise to comprehend its functioning, eliminate brain diseases, and ultimately upload ourselves, excited the public imagination with its near-religious imagery. As those initial simulations proved to be computationally underpowered and inaccurate, this promise backfired, leading to the withdrawal of public support for some time in the 2020s. Much was learned, but the public was disappointed.

Paraphrasing the twentieth-century British war leader Winston Churchill, neuroscience was at the end of the beginning of the quest to understand the brain and the mind. Neuroscientists had not yet figured out how to bridge the many levels of neurophysiology, from molecules to cells to circuits to behavior, but they had discerned enough to make the mission clear, and many critical tools were in place.

The Modern Era: 2064

Today, by identifying hierarchies of modules and submodules in the cortical sheet, we've largely tamed the sheer diversity and the vast extent of the neocortex. The basic organization of the cortical six-layered sheet is now known to schoolchildren, and if the overall interconnectivity is far too hard for any individual to understand, the nervous system of laboratory organisms like flies can now be emulated—successfully—with computers; human brains, too, have been simulated with some fidelity, although in time frames—about one-hundredth of real time—that make them less useful than was originally anticipated.

The retina was the first piece of neural tissue to be understood, in the sense that its output—action potentials along the optic nerve—can be quite accurately predicted from its input—patterns of light. One reason the retina led the way is its (relative) simplicity; unlike other nervous matter, the retina has primarily feed-forward connections—without any significant connections from the brain proper back to the retina. Most of its cellular elements had been recognized in the late twentieth century. By 2020, a Big-Science consortium of anatomists, physiologists, biophysics modelers, and machine learning specialists had arrived at a nearly complete description of retinal input-output, and the firing rates of the two dozen ganglion cell types, whose axons make up the optic nerve, could be reliably predicted, in response to arbitrary visual stimuli. That understanding (in combination with advanced optogenetics and implantable ocular electronics) led to effective treatments for macular degeneration, diabetic retinopathy, and retinitis pigmentosa.

Similar techniques helped crack the codes used in the visual thalamus and early visual cortical areas, as the onion layers of the brain began to be peeled back, one by one. A complete cellular-based working model of how the mouse moves through a maze in response to what it sees, together with the ontology of the approximately one thousand different cell types that make up the brain, was achieved in the mid 2020s. The senses of touch, hearing, and smell were decrypted a few years later.

This success fed the hope that understanding the entire mouse brain could not be far behind. Mechanistic explanations for what happens when the brain goes to sleep, dreams, wakes up, decides to run, remembers a location for another day, and develops across its lifespan, from birth to senescence, seemed close at hand. But these hopes were dashed. Yes, plenty of individual stories were told, but they could not be assembled into a coherent whole.

Funding for brain research slowed down because of the inability to translate these insights to people and their pathologies. Not that anybody seriously argued that the human brain was fundamentally different from that of the mouse. Of course, the two differ dramatically in size and accessibility. The human brain is more than a thousand times bigger than the mouse brain—1.4 kg versus 0.4 g in mass; a papaya versus a sugar cube in volume; eighty-six billion nerve cells versus seventy-one million for the entire brain and sixteen billion versus fourteen million nerve cells for the neocortex. Even more importantly was the ethical constraint: the living human brain could only be probed at the required cellular level under rare conditions, primarily during neurosurgery. fMRI, EEG, MEG, and other noninvasive techniques that peered at the brain from the outside were blind to genes, proteins, and cell types. While a rice- or corn-sized chunk of human gray matter is by and large similar to that of the mouse, there are many, many minute differences. Given the divergent ways in which *Mus musculus* and *Homo sapiens* evolved over the last seventy-five million years since their last common ancestor, their genes and gene regulatory mechanisms, proteins, synapses, neurons, and circuits differ in a multitude of small ways. Yet these trivial but elusive differences made generalizations from the mouse to humans difficult. Indeed, pharmaceutical companies had realized this earlier on and had discontinued much of their mouse research already in the early 2010s. After the animal rights movement managed to shut down almost all invasive research on nonhuman primates worldwide by the end of the 2020s, neuroscience entered what is now known as the lost decade. This was marked by low funding and pessimism that neuroscience could ever truly ameliorate the staggering toll that brain diseases took on the aging population, estimated to be 10 percent of world GDP.

The darkest hour is often just before the dawn. Help came from a very distant relative of humans, *C. elegans*, and from the triumphant marriage of artificial and biological molecular machines.

To be sure, it took over thirty-five years from when the connectome of two worms were mapped (in 1986) for an accurate, predictive, comprehensive, and fully testable model of its nervous system to be formulated. The key insight—the role of neuromodulators in switching pathways and circuits dynamically—was already faintly recognized fifty years ago by such pioneers as Cornelia Bargmann and Eve Marder, in the worm and other non-vertebrate species, but because worms lack action potentials, the importance of Bargmann and Marder's work for

vertebrate creatures was initially overlooked. We now know that principles of dynamic routing are critical in all creatures.

The conquest of the living human brain was finally achieved with
nanobotic neural implants, colloquially known as brainbots. These are
molecular machines for imaging and manipulating the brain that can
be safely injected by the millions into the bloodstream. The first generation of brainbots were designed to sample and measure their local
environment, such as the electrical potential, or the concentration of
a particular neurotransmitter or small molecule, and could be queried
from the outside. More advanced probes read the transcriptional sig-
nature of individual neurons, monitor their electrical activity, arrest or
trigger spikes, and, most recently, control synaptic release at individual
synapses. They intervene at any point in the body by delivering missing
or eliminating miss-formed neurotransmitters or proteins, or trigger
electrical activity. Some operate transiently while others act as modified viruses that find a permanent home inside nerve and glial cells to
arrest and ultimately repair the damages degenerative diseases such as
Alzheimer's or Parkinson's cause; by the mid 2050s, almost all medicine,
and all neuroscience, had moved to nanobotic platforms; even optogenetics, the workhorse of the early twenty-first century, eventually was
displaced. Because of their high spatial specificity—guided by an externally imposed 3-D radio field—properly designed nanobots can target
individual cells anywhere in the brain with enormous precision.

Many once-common mental diseases can now be delayed or, in a few
cases, cured. To be sure, progress in reducing morbidity and mortality
of brain-based pathologies—tumors, traumatic-brain injury, epilepsy,
schizophrenia, Parkinson's, Alzheimer's and other forms of dementia—
took much longer to realize than anyone conceived of in the early years
of the new millennium. (An instructive parallel is the War on Cancer,
announced by President Nixon in 1971, when America was flush with
the success of the lunar landing; it was nearly five decades before there
was a significant decline in the actual death rates for cancer, while death
rates for respiratory, infectious, and cardiovascular diseases had plum-
meted much earlier.) Reducing the collective impact of brain-based pathologies turned out to be *more* difficult than curing the diverse set of
pathologies known as cancer; both are highly heterogeneous diseases

with an inexhaustible multiplicity of genetic, epigenetic, and environmental causes, but because of mosaicity, the complexity was even greater for the brain.

Brainbot treatment is expensive. And like most medical procedures, it has side effects, restricting it to appropriate patient populations. Yet although traditionalists and religious people object, nanobotic enhancement in healthy subjects is immensely attractive to those who believe in the infinite betterment of the human condition. Its proven ability to boost athletic agility and speed, learning and recall, has given rise to an underground market in brain enhancements. Those able to pay and willing to live with the short- and long-term morbidity and mortality risks are threatening to turn into trans-humans, a cognitive elite that easily outcompetes nonenhanced normals in the marketplace and in warfare.

In academic circles, the ongoing debate is about the growing raft of whole-brain simulations and what they mean both ethically and scientifically. For one thing, the question—first raised over fifty years ago—about the relevant level for brain simulation lingers. The intellectual tension arises between bottom-up simulators, who hold a form of extreme biological chauvinism—the need to consider every ionic channel, synapse, and action potential to fully do justice to the baroque complexity of the brain's circuits—and top-down simulators, who are motivated by the austerity of a purely algorithmic approach of replicating the mind in software (the mind is not wet, after all) and start with behavior or with computation.

Both sides have made major advances, but neither has been fully successful. Biophysicists accurately simulate the biochemical and neural activities of worms and flies with near full verisimilitude. Yet for mammals, deviations appear. And these differences between actual and simulated behaviors become more pronounced when moving from rodent brains, via those of monkeys and apes, to the human brain. Thus the spoken language such simulations produce is garbled, and most simulations remain at the kindergarten level on many tasks. What are we missing today? Do we have to simulate every ionic channel and every neurotransmitter molecule? Must we treat the brain as a quantum mechanical system? The brain is, after all, a physical object like any other one, subject to the iron law of quantum mechanics. Yet the vast majority of brain scientists assume that the nervous system, a hot (by QM standards) and wet organ closely coupled to its environment, can be approximated very well as a classical system.

Even considered as a classical system, biophysical brain simulations are dreadfully slow, working at one-hundredth the speed of real human brains; now that Moore's law has run out, and quantum computation proved to be of limited real-world use, it's not clear where the next advance will come from. Top-down modelers, meanwhile, capture some of the essence of human cognition, but with comparatively little fidelity to biological reality. Until the two approaches can be bridged, the thought-reading prosthetics that seemed so near a decade ago will continue to remain elusive. (In part, once again, the problem stems from complexity. Mathematicians and engineers imagined that there would be one true brain algorithm to rule them all, but because of the arbitrary accidents of nature's evolutionary opportunism, that simply hasn't proven to be the case; indeed, there seem to be almost as many algorithms as there are brain circuits, which has left little opportunity for shortcuts along the way.)

Meanwhile, on the cognitive side, processes such as language, planning, social cognition, and higher-level reasoning still resist explanation, especially in the intricate forms they take in people. Nanobotics may bridge this gap in our knowledge eventually, but for now, knowledge of uniquely human faculties still lags. We still don't know how the brain encodes sentences, and only a tiny bit is known about word meanings; complex concepts, like "the sort of person who reads fictitious narratives," remain entirely out of our grasp. If the neural basis of association has been entirely unraveled, the neural basis of higher-level cognition has not.

Ethically, as full-scale human brain emulations have neared, the political battles have been heated. Some see modeled rodents as ethically equal to real rodents and argue that complete human-brain emulations merit rights equal to human beings. Some scholars see emotional distress in the rudimentary human brain simulants. Yet most (chose to) believe that a simulation is an imitation rather than the real thing, just like a computer simulating the aerodynamics of flight will never actually lift off. Politicians avoid the issue, but time is clearly running out. Will it be legal to employ a whole-brain emulation for intellectual work, much as one might employ a human? Would it be ethical? Does all income accrue to the owner of the simulation, or might those whose brains contributed to the simulation also deserve royalty fees, in addition to the hourly fees they were paid for their original participation in extended brain scans?

The final challenge, indubitably, will be how *subjective* feelings, how consciousness itself, emerges from the physical brain. Even today, there remains an explanatory gap between neural activity and subjective feelings, between the brain and the conscious mind. One belongs to the realm of physics, to space and time, energy and mass. The other one belongs to a still poorly understood magisterium of experience. Spearheaded by the molecular biologist turned neuroscientist Francis Crick, cognitive neuroscientists have been tracking down the neuronal correlates of consciousness, but the vast complexity involved has kept us from a full solution. If we by now have a clear understanding of the dynamics by which information passes into awareness, we still don't fully know why experiences feel the way they do. The expectation is that the "hard problem" of consciousness will eventually be dissolved, and even disappear, much in the same way that the problem of "what is life" has disappeared from view, replaced by a host of more tractable problems about the details of reproduction and metabolism. As the behavior of computer artifacts begins to approach, and often to exceed, human capability, more and more people believe that consciousness arises from a privileged form of information associated with highly organized matter, such as brains or artificial intelligence agents, as argued already half a century earlier by Giulio Tononi. But if Descartes's famous conclusions four centuries ago might be paraphrased "I am conscious, therefore I am," the issues of consciousness still haven't been fully resolved. It is to be hoped that the next hundred years will finally bring resolution to the ancient mind-body riddle.

Acknowledgment: We wish to thank Ramez Naam, author of *Nexus*, for very thoughtful comments.

GLOSSARY

Action potentials. A rapid event in which the electrical membrane potential of a cell rises (or depolarizes) and then falls (hyperpolarizes), due to the opening and closing of ion channels. An action potential typically occurs due to sufficient neurotransmitter release from presynaptic neurons, and itself elicits neurotransmitter release at the axon terminal, which elicits depolarizations in postsynaptic targets. In this way, action potentials are the primary means by which neurons communicate with one another.

Channelrhodopsins. A special family of proteins that act as light-gated ion channels; they open when exposed to light. Naturally occurring in unicellular green algae, these proteins can be expressed in neurons through genetic transfection. Because ion channel opening triggers depolarization, channelrhodopsins can be used to artificially stimulate neurons with light.

Cre driver line. A genetically engineered breed of laboratory mice that allows scientists to regulate genes in very specific subpopulations of cells at a particular developmental time point. The most popular technique uses the Cre-loxP system to target cells at a variety of spatiotemporal scales, from ubiquitous expression throughout the adult mouse to only expressing in a molecularly characterized subset of excitatory or inhibitory cortical cells. In combination with Cre reporters, these molecularly characterized cells in these Cre line animals can be made to be fluorescent or can be turned on or off with different colored beams of lights or drugs (opto- or pharmacogenetics).

Cytoarchitecture. The study of the cellular composition and structure of the brain's tissues using the microscope.

Diffusion MRI. An MRI-based technique that measures diffusion of water molecules in biological tissues, with primary application in studying fiber structure and connectivity in the brain.

Diffusion tractography. A 3D modeling technique used to visually represent neural connections determined by diffusion of water motion along axon tracts in the brain.

DNA bar code. An arbitrary string of DNA letters, used to identify a molecule, cell, or other entity.

Electroencephalography (EEG). A technique of recording electrical activity along the scalp of the head by the placement of many electrodes that are cross-calibrated. In addition to its use in basic science, the technique has been applied in diagnosis of epilepsy and other brain disorders.

Electrophoresis. The motion of dispersed particles in a fluid by applying a uniform electric field. Often applied to identify or quantify segments of RNA, DNA, or proteins.

Exome. The part of the DNA that is actually transcribed into RNA, about 1% of the human genome.

Fluorescent In Situ Sequencing (FISSEQ). the process of reading the sequence of letters along a DNA strand in the context of an intact slice of tissue, by using an automated microscope.

Functional magnetic resonance imaging (fMRI). An application of MRI technology that measures brain activity by detecting associated changes in blood flow to measure neuron activation.

Gene expression. The process by which information in DNA is synthesized through RNA into proteins. All known life forms use it.

Halorhodopsins. Like channelrhodopsins, these proteins are light-gated ion channels that can be made to express in the membranes of neurons, but these channels specifically transmit chloride. When stimulated with light, the opening of these channels causes hyper polarization, which suppresses neuronal responses. Combined with channelrhodopsins, these channels provide a means to both turn off and on neural activity with light.

Histology. The study of microscopic anatomy of cells and tissues, using various stains for cells and tissues as well as expert diagnostics.

Immune microscopy. Using special forms of molecular recognition called antibodies (derived from the immune system) to tag particular proteins or other molecules with colored dyes or with DNA barcodes for visualization by in situ sequencing or in situ microscopy.

In situ hybridization. A gene expression detection technique in which a single probe for each gene is designed and hybridized to RNA in intact tissue retaining the spatial context.

Light-sheet microscopy. A technique for microscopy, typically in living organisms, in which a sheet of laser light illuminates a thin section of tissue, produce sharp high-contrast images with relatively little interference from the non-illuminated tissue.

Magnetoencephalography (MEG). A functional neuroimaging technique that measures changes in magnetic fields in the brain to study cognitive processes and clinical changes.

Magnetic resonance imaging (MRI). A medical imaging technique using powerful magnets that applies the nuclear resonance properties of atoms to create detailed images of the body.

Microarray. An array containing thousands of small DNA or RNA sequence probes that can perform genetic tests by applying an independent tissue using imaging.

Optogenetics. A technique for using light to control neurons. See channelrhodopsins and halorhodopsins.

Positron emission tomography (PET). A medical imaging technique that detects gamma rays emitted from an injected radioactive tracer in the body. PET produces a three-dimensional image of functional activity in the brain.

Sequence space. The enormously large abstract set of all possible DNA bar codes.

Single gene disorder. A disorder caused by a mutation in a single gene.

Two-photon microscopy. A fluorescent imaging technique that allows high-resolution imaging of living tissue to a depth of about 1 mm.

INDEX